These articles are selected to suit the vocabulary and experiences of sharp high school students who are interested in Science. Students who pass Advanced Placement-level courses should "get" this book.

If you're not sure whether an article is "true", assume that it isn't, and that the author is clever.

For Lumin and Mason, who are great at laughter as well as Science.

A great gift for scientists, doctors, engineers, and science teachers is a subscription to **The Journal of Irreproducible Results**®, founded in 1955 to make scientists and doctors laugh. Subscribe at www.jir.com.

Cover by Susan Newman, www.susannewmandesign.com.
Back cover photo of Norman Sperling by Richard Berry.

Table of Contents

v

The Definitive Determination of How Many Licks It Takes to Get to the Center of a Tootsie Roll® Pop

Candice Caine, Colorado Community College of Candies, Cakes, Cookies, and Confectionaries

Abstract

We present the results of a scientific study for determining the number of licks it takes to get to the center of a Tootsie Roll Pop. In addition to utilizing numerous individuals who each lick multiple lollipops, we introduce a new metric developed for resolving the number of licks, which takes into account the largest variation among individual lickers, namely the surface area of the tongue.

KEYWORDS: CENTER; DETERMINE; LICKS; LOLLIPOP; TOOTSIE ROLL.

Introduction

In 1970, Tootsie Roll Industries aired one of the most famous television commercials of all time, featuring a young boy attempting to ascertain how many licks it takes to get to the center of a Tootsie Roll Pop. First, he asked Mr. Turtle. But Mr. Turtle didn't know, so he suggested the boy seek the wise counsel of Mr. Owl. After locating the bespectacled raptor, Mr. Owl replied, "Let's find out." He proceeded to unwrap the lollipop in question and place it firmly within his mandible. Mr. Owl licked it once, then twice, and just as he was about to take a third lick, he impatiently bit into the Tootsie Pop, and blurted out, "Ahem ... 3."

Ever since then, the scientific community has been baffled and bewildered by this seemingly unfathomable mystery. In fact, many nay-sayers have predicted that the world may never know.

With this in mind, we decided to attempt to solve this perplexing problem ourselves. First, we embarked upon an exhaustive literature survey to see if there were any previous investigations that had reported significant progress in resolving this elusive question. We found only one legitimate publication that lay claim to such an accomplishment [1]. However, after carefully reading through the article, we were disappointed to learn that the researchers had only performed their experiment using a solitary subject licking a single lollipop, thus preventing any meaningful statistical analysis.

In response, we developed our own statistically-sound scientific study, which used a wide variety of subjects who each licked multiple lollipops. This large sample of data enabled us to accurately determine the mean value of the number of licks it takes to get to the center of a Tootsie Roll Pop, along with an accompanying statement of uncertainty. We describe our experiment in detail, discuss our findings, and introduce a new metric developed for resolving the number of licks, which takes into account the largest variation among individual lickers, namely the surface area of the tongue.

The 8 Tootsie Roll Pops that were distributed to each of the 10 participants.

Methodology and Findings

Our first step was to find 10 willing participants, representing both sexes, of various ages. In addition to myself, I found 9 other enthusiastic volunteers, including my husband, Russell Stover; my 8-year-old twin sons, Mike and Ike; my 3-year-old daughter, Baby Ruth; my father, Grandpa O'Henry; 2 next-door neighbors, Mr. Goodbar and Mrs. Fields; and 2 of my graduate students, Ted Nougat and Reese's Witherspoon.

Each participant was given 8 Tootsie Roll Pops – 2 of each flavor (orange, cherry, grape, and chocolate), as shown in the picture. Along with the lollipops, each subject was provided with explicit instructions to count the number of licks required to get to the center of each one. Everybody had one week to perform their licking experiments. Afterwards, I collected the data from the participants and computed mean values and standard deviations for each of the individuals.

Unfortunately, data from 2 of the subjects had to be eliminated. Like Mr. Owl, my son Mike was too impatient to ever make it to the center without first biting. And since Grandpa O'Henry informed me that he kept falling asleep with the lollipops in his mouth, we had no choice but to toss his skewed data into the trash.

The remaining 8 data sets are plotted in Figure 1, along with their respective standard deviations. Upon examining the data, we were surprised at how large the differences were among the participants as compared to the individual standard deviations. For example, mean licks fluctuated all the way from 50 to 475, while the largest individual standard deviation was only 40. Statisticians describe results such as these as being inconsistent [2]. In this type of situation, the scientist is usually advised to look for any unnoticed systematic errors. So that is exactly what we did, and after much deliberation, we came to the conclusion that the large discrepancy appeared to be due to the widely divergent surface areas of the participants' tongues. For instance, Ike and Baby Ruth, the youngest of the volunteers, required the most licks and have the smallest tongues, while Ted Nougat required the fewest licks by far and has a tongue nearly as large as Gene Simmons's.

Because of this, we developed a new metric that we refer to as 'normalized licks.' This new quantity is calculated by multiplying the 'raw' licks by the individual's surface area of the tongue, and then dividing by the population's average surface area of the tongue. The equation is shown below as:

$$\text{Normalized Licks} = \frac{\text{Licks} \times \text{Surface Area of Tongue}}{\text{Average Surface Area of Tongue}}$$

3

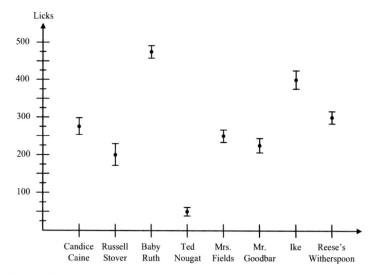

Figure 1. Mean values and standard deviations of the number of licks for each of the 8 valid participants.

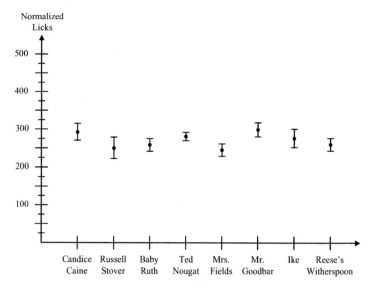

Figure 2. Mean values and standard deviations of the number of normalized licks for each of the 8 valid participants.

We calculated the 'normalized licks' for each of the participants, and they are plotted in Figure 2. Upon examining this new set of data, we see that the differences among the participants are now on the same order as the individual standard deviations.

Since the 'normalized licks' are consistent, we can now find the best estimate based on these measurements using a weighted average, where the weights are the inverse squares of the corresponding uncertainties, assuming of course that our measured licks were governed by the usual Gaussian distribution. Likewise, the overall uncertainty can be easily calculated using error propagation. After performing the aforementioned computations, we have determined that the number of 'normalized licks' it takes to get to the center of a Tootsie Roll Pop is 273 ± 8.

Conclusion

We have described a method for determining the number of licks it takes to get to the center of a Tootsie Roll Pop. Rather than simply combining the measured licks, we found a more accurate representation to be a newly developed quantity that we refer to as 'normalized licks,' which takes into account the largest variation among individual lickers, namely the surface area of the tongue. We found the best estimate for the number of 'normalized licks' to be between 265 and 281.
We speculate that the remaining uncertainties can be divided into 2 categories, those relating to the individual lickers and those relating to the lollipops themselves. Possible variations in the lickers may include such attributes as pH levels in the saliva, densities of the papilla (taste buds) on the tongue, pressure applied by the tongue, and possible rotations of the lollipop between licks. Inconsistencies in the lollipops may include age, color, and how accurately the chewy Tootsie Roll is centered within the hardened candy shell.

References

[1] Cadbury, Ghirardelli, and Nestlé, "Licks of the Tongue Required to Arrive at the Chewy Core of a Tootsie Roll Brand Lollipop", *Journal of Applied Confectionary*, vol. 6, no. 9, pp. 457-783, April 1975.

[2] J. R. Taylor, "An Introduction to Error Analysis," University Science Books, 1982.

About the Author

Candice Caine was born in Hershey, Pennsylvania, in 1967. She received her BA degree in 1990 and her PhD in 1994, both in Home Economics from Texas M&M. In 1995, Dr. Caine was hired by Brach's, Inc., where she developed the world's first genetically modified candy corn. Unfortunately, she was laid off in 2002 during a corporate fun-sizing. In 2003, Dr. Caine joined the faculty at the Colorado Community College of Candies, Cakes, Cookies, and Confectionaries (C.C.C.C.C.C.C.), where she performs research and teaches graduate-level baking classes. Dr. Caine has authored or co-authored over 1,000 technical papers and 23 books, including the worldwide bestseller, *Cotton Candy for the Soul.*

Please send any correspondence to Jeffrey Jargon at: thejargons@aol.com.

Nature Versus Nurture: One Man's Diabolical Experiment on His Own Children

Richard Dreison, Broomfield, Colorado

Abstract

I present the results of an 18-year-long case study that I performed on my 3 sons, who are identical triplets. The purpose of this research was to attempt to understand the hereditary and environmental influences on behavioral development, while at the same time freeing me from raising all 3 children on my own.

Introduction and Motivation

"Nature versus nurture" refers to the relative importance of innate qualities, as opposed to personal experiences, in determining an individual's physical and behavioral traits. One method scientists have used to determine the contributions of genes and environment is to study identical twins raised apart [1-5].

Here, I take the concept a step further by studying identical triplets. How I originally came up with the idea of performing such a groundbreaking case study was actually completely unintended. You see, 18 years ago, my wife, Julia Louis-Dreison, was pregnant with triplets. Her pregnancy was progressing normally, and we were quite excited, although admittedly a bit apprehensive, at the prospect of caring for triplets, seeing as we were inexperienced, first-time parents.

Julia originally intended a natural childbirth. In fact, she took a Lamaze class during the pregnancy to learn breathing techniques that would help her gain "mind over matter." But shortly after we arrived at the hospital, the pain became too overwhelming, and she decided to go with the "drugs over matter" approach and have an epidural. Unfortunately, instead of injecting her with a regional anesthetic, she was mistakenly given some sort of neurotoxin. Julia immediately went into cardiac arrest and died within a few minutes, despite repeated attempts by the doctors to resuscitate her.

Luckily, they were able to perform an emergency C-section and save the babies. I have to admit, it was a very emotional day for me; within a 2-hour period, not only I had lost my dear wife, but I had also witnessed the birth of my 3 sons.

Despite the tragedy of my wife's untimely death, I was at least grateful in knowing that I didn't have any undue financial burdens. Previously, I had taken out a $1,000,000 life insurance policy on Julia. And on top of that, I won a $2,000,000 malpractice suit against the hospital. Despite my newfound wealth, I soon realized that it would be difficult, if not impossible, for me to raise the triplets on my own. I figured I could probably manage one, but 3 would definitely exceed my limits. I spent several days trying to come up with a feasible solution, and then one morning I suddenly awoke with a brilliant plan – I would perform the ultimate nature versus nurture experiment by having my children brought up in radically different environments. One would be raised to become a warrior, one a pacifist, and the other I would bring up on my own as a control subject. In the following section I describe the experiment in detail, and discuss my findings.

Methodology and Findings

Shortly after coming up with my plan, I gave my children names that would reflect their respective upbringings. I named the warrior-to-be Genghis; the pacifist-to-be, Gandhi; and the control child, Greg. Next, I tried to determine where I could send Genghis and Gandhi to carry out the experiment. After an exhaustive search, I found 2 institutes that would bring them up in drastically opposing circumstances.

I would enroll Genghis at St. Michael's Christian Military Academy in Boise, Idaho. St. Michael's is an all-boys preparatory school that teaches military tactics in conjunction with a literal interpretation of the Bible. When the headmaster, Reverend General "Stormin', Reformin'" Norman Foreman, told me that the school motto is "We kill more heretics before Easter than most Christians do all year," I knew this was the right place for Genghis.

Next, I would enroll Gandhi at the Southwestern Hindu Institute of Tranquility, despite its rather unfortunate acronym. Located in Las Vedas, Nevada, this boarding school's curriculum includes Hindu principles and history, yoga, spiritual enlightenment, and call center training.

As for Greg, I would raise him myself, enrolling him in public schools and encouraging him to embrace mediocrity. To achieve this, I would enforce a strict adherence to the mean and persuade him to do just enough studying to remain a straight C student. Furthermore, I would allow no extracurricular activities on his part aside from loitering at the mall with his friends.

Once Genghis and Gandhi were sent their separate ways, I decided that if the experiment were to work, I could not let any of the 3 know that the other 2 existed. I did, however, visit Genghis and Gandhi from time to time to keep tabs on them, but I did it under the guise of business trips so that Greg wouldn't suspect anything.

I kept this case study going and took copious notes on their progress for 18 years, at which time I decided to finally conclude the experiment and introduce the 3 brothers to one another. But first I had each of them fill out a survey I devised to reveal key personality traits. The table summarizes their major differences. One can clearly see that their wildly divergent upbringings greatly influenced their preferences in a wide number of subjects, including religious beliefs, hobbies, major accomplishments, and favorite *Star Wars* characters, to name just a few.

Despite their numerous distinctions, however, Genghis, Gandhi, and Greg do share a fair amount in common as well. Aside from obvious similarities in appearance, height, weight, and initial signs of male pattern baldness, all 3 wear boxers instead of briefs, are afraid of clowns, enjoy the comic musings of Patton Oswalt, exclusively use the Garamond font in all written correspondence, and are allergic to peaches, poodles, and polyester. Additionally, all 3 list blue as their favorite color, although each one prefers a different shade – Genghis favors Navy blue, Gandhi fancies sky blue, and Greg likes plain old blue.

As you can imagine, my 3 identical sons were quite taken aback when they were first introduced to one another. And at that moment, I learned that they also had one more trait in common, namely their anger with me for separating them at birth. However, they each

voiced their disapproval in their own unique way –
Genghis threw a punch, Greg threw a tantrum, and
Gandhi threw up.

Despite their contrasting backgrounds, the 3 of
them have remained inseparable since their initial
reunion. In fact, each one has expressed a sincere
interest in the others' activities and habits. For
instance, Genghis has developed a taste for vegetable
curry, and has learned the joy of TiVo®; Gandhi has
developed a taste for red meat, and has increased his
number of earthly possessions to 4 by purchasing an
iPod®; and Greg now wakes up before noon, and enjoys
the great outdoors, whether it be watching birds with
Gandhi, or hunting them with Genghis.

Conclusion

In this paper, I have summarized an experiment
that I performed on my offspring. Taking identical
triplets with essentially the same genes, I separated
them at birth and observed their differences and
similarities as they developed. With such a vast array of
documented distinctions and commonalities among
Genghis, Gandhi, and Greg, I'm afraid that I have no
conclusive assertions other than to say that both nature
and nurture appear to play vital roles in the upbringing
of children.

References

[1] Cherny, S. S., DeFries, J. C., & Fulker, D. W. (1992). Multiple
regression analysis of twin data: A model-fitting approach. *Behavior
Genetics, 22,* 489-497.
[2] Robinson, J. L., Kagan, J., Reznick, J. S., & Corley, R. (1992).
The heritability of inhibited behavior: A twin study. *Developmental
Psychology, 28,* 1030-1037.
[3] Zahn-Waxler, C., Robinson, J. L., & Emde, R. N. (1992). The
development of empathy in twins. *Developmental Psychology, 28,*
1038-1047.
[4] Reznick, J. S. (1997). Intelligence, language, nature, and nurture
in young twins. In R. J. Sternberg & E. L. Grigorenko (Eds.),
Intelligence: Heredity and Environment. New York: Cambridge
University Press.
[5] Reznick, J.S. & Corley, R. (1999). What twins can tell us about
the development of intelligence: A case study. In M. Anderson
(Ed.), *The Development of Intelligence.* London: University College
Press.

Question	Genghis	Gandhi	Greg
Favorite Hobby:	Hunting Endangered Animals	Volunteering at a Homeless Shelter	Watching TV
Favorite Book:	*The New Testament*	*The Bhagavad Gita*	*Harry Potter and the Half-Blood Prince*
Favorite Movie:	*Lethal Weapon 4, The Passion of the Christ* (tie)	*Gandhi, The Razor's Edge* (tie)	*Harry Potter & the Prisoner of Azkaban*
Favorite TV Show:	*Walker, Texas Ranger*	*Wai Lana Yoga*	*Joey*
Favorite Card Game:	Texas Hold 'Em	Solitaire	Go Fish
Favorite Musician:	Nine Inch Nails	Enya	'N Sync
Favorite Sport:	Ultimate Fighting	Yoga	Football
Favorite *Star Wars* Character:	Darth Sidious	Yoda	Jar Jar Binks
Favorite Food:	Blood Sausage	Yogurt	Burger & Fries
Favorite Beverage:	Red Bull	Chamomile Tea	Diet Coke
Religious Preference:	Evangelical Christian	Hindu	Unitarian
Pets:	Tarantula, Scorpion, Python	None	Spot the Dog
Pet Peeves:	Blasphemy	Licorice-Scented Incense	Dad's Refusal to Get Satellite TV
Strengths:	Weapons Proficiency, High Pain Threshold	Compassion, Empathy	Conformity, High Metabolism
Weaknesses:	Hair Trigger Temper, Women in Uniform	Generous to a Fault, My 3 Earthly Possessions	Procrastination, Sleeping Late
Biggest Accomplishment:	Memorizing the *Bible* at Age 10	Achieving Enlightenment at Age 10	Seeing Every Episode of *Friends*
Greatest Fear:	Being Captured Alive	Being Reincarnated as a Dung Beetle	Cable TV Going on the Blink
Name 3 People You Would Invite for Dinner (Dead or Alive):	Gen. George Patton, Gen. Norman Schwarzkopf, Gen. Robert E. Lee	Mahatma Gandhi, Martin Luther King, Jr., The Dalai Lama	Ben Affleck, Simon Cowell, Jeff Probst

American π

Lawrence M. Lesser, PhD, University of Texas at El Paso

☛ Goes well to the tune of Don McLean's *American Pie*

CHORUS: Find, find the value of π,
Starts 3 point 1 4 1 5 9.
Good ol' boys gave it a try,
But the decimal never dies,
The decimal never dies ...

In the Hebrew Bible we do see
the circle ratio appears as 3,
And the Rhind Papyrus does report
4/3 to the fourth,
& 22 sevenths Archimedes found
with polygons was a good upper bound.
The Chinese got it really keen:
3-5-5 over one-thirteen!
More joined the action
with arctan series and continued fractions.
In the seventeen-hundreds, my oh my,
the English coined the symbol π,
Then Lambert showed it was a lie
to look for rational π.
He started singing ... (REPEAT CHORUS)

Late eighteen-hundreds, Lindemann shared
why a circle can't be squared
But there's no tellin' some people —
can't pop their bubble with Buffon's needle,
Like the country doctor who sought renown
from a new "truth" he thought he found.
The Indiana Senate floor
read his bill that made π 4.
That bill got through the House
with a vote unanimous!
But in the end the statesmen sighed,
"It's not for us to decide,"
So the bill was left to die
Like the quest for rational π.

They started singing ... (REPEAT CHORUS)

That doctor's π-in-the-sky dreams
may not look so extreme
If you take a look back:
math'maticians long thought that
Deductive systems could be complete
and there was one true geometry.
Now in these computer times,
we test the best machines to find
π to a trillion places
that so far lack pattern's traces.
It's great when we can truly see
math as human history —
That adds curiosity ...
easy as π!
Let's all try singing ... (REPEAT CHORUS)

EXPRESSIONS OF MAGNITUDE: A RUNYON-LIKERT SCALE

Louis G. Lippman, PhD, Western Washington University

Abstract

Damon Runyon's writings include a number of characteristic expressions that are meant to convey differences in magnitude. Those expressions were aligned with more conventional values on Likert scaling, with the result that a series of descriptors was developed that can be used for scaling more subjective perceptions of degree or amount.

Use of the remarkably versatile "Likert-type" rating has become extremely common. "Likert-type" rating scales put contrasting terms on either end (often "agree" and "disagree"), and a scale of numbers, with descriptors, in between.

Thousands of professionals undoubtedly made use of some variant on their theses and dissertations, and hundreds of thousands of undergraduates, performing research as a requirement in methods classes, have made use of Likert-type scaling. It would be astonishing if more than a handful of these individuals actually referred to the seminal monograph from June, 1932.[1] It would be even more amazing if any of these people had ever *read* the report, and even more unlikely if any of these assiduous investigators would have any idea of Likert's first name.[2]

A contemporary of Likert, and fellow New Yorker, was Damon Runyon, a talented newspaperman who is best known for extraordinarily inventive stories about a variety of shady characters on New York's Broadway. His writing is unique and infectious. In contrast to American Psychological Association style, everything is in present tense – even when the narrator (who is never identified) recounts previous events.

The stories are loaded with creative slang, and formed the basis for the musical *Guys and Dolls*. A particularly striking feature of Runyon's writing is his inventive use of adjectives and adjectival phrases for suggesting different magnitudes.

Regrettably, these 2 New Yorkers evidently never met or collaborated professionally. The purpose of the present effort is to compensate for this unfortunate omission. Hence, it is suggested that by pooling the most creative aspects of both of these individuals, the research community could have an even more useful and incisive scaling tool.

Method

Semi-systematic scanning through collections of Runyon's short stories[3] yielded 20 adjectives and adjectival phrases which, in context, had been used to suggest different magnitudes. These items were printed in alphabetical order; next to each was a blank in which participants were asked to supply a rating. The printed form also summarized instructions.

Raters were the 11 Writing Fellows and the one director from the 1999-2000 program at Western Washington University. Writing Fellows are undergraduate seniors, across disciplines, who had demonstrated exceptional proficiency and thus had been nominated for a special program that would train them to assist in writing-intensive courses, and to serve as consultants to students. It was assumed that these individuals would have particular sensitivity for words and word meanings.

These 12 individuals were instructed to scan through the entire list of adjectives, try each in sentences, and rate the quantity that each seemed to imply or connote, using a scale ranging from 1 to 7, where the only anchors were "1" to indicate the least and "7" the most.

Results and Discussion

Mean ratings showed that Runyon's adjectives did span a wide range of suggested magnitude. On the basis of those data, the following 9-point Runyon-Likert scale was constructed:

> 9: More than anything else
> 8: Very much indeed
> 7: Quite some
> 6: Some greatly

5: No little
4: More than somewhat
3: Somewhat
2: Some little
1: Not much

This derived scale holds great potential whenever there is need to have raters express subjective reactions or the implicit and suggested attributes of stimuli. As such, it should prove to be a valuable research tool that could serve the next generation of investigators as nicely as the original Likert development has served the last 3 generations of scholars.

References

1. Likert, R. (1932) A technique for the measurement of attitudes. *Archives of Psychology,* No. 140.
2. Rensis
3. Runyon, D. (1950) *Runyon on Broadway.* London: Constable. (1954) *Runyon from first to last.* London: Constable.

Adapted, with permission, from *Psychology and Education – an Interdisciplinary Journal,* vol. 42, no. 1, 2005, pp 38-39.

☛ **What other interesting Likert-type scales can you concoct?**

Experiments Based on Commonly Held But
Seldom Tested Beliefs, Part I:

A Saucepan Under Observation Never Reaches a Phase Transition

Iva P. Aitchdee, Underwood Liberal Arts
College, Box Springs, Colorado

In this first experiment I tested that oft-repeated bit of kitchen lore — "a watched pot never boils". The assumption is made that this statement refers to a common kitchen pot on an ordinary stove filled with a liquid typically used for cooking.

To perform this experiment I took a small saucepan, filled it with ordinary tap water, and placed it on the stove. Then I pulled up a chair and sat down to faithfully watch. Soon the water began to boil, and after a substantial amount of time it had all changed to its gaseous state, but true to the old maxim, the pot itself remained intact.

After the excitement with the water boiling away was over I soon grew tired of watching the pot. "Never" is a rather stiff requirement, and I was not looking forward to sitting in the kitchen of my apartment forever to prove that this age-old theory was, in fact, true.

To solve this little problem I looked up the boiling point of both aluminum and steel, the materials that were listed on the bottom of the pot. Both were far above the temperature that my stove could reach. Therefore, due to the laws of thermodynamics that state, simply, that things can't get any hotter than you heat them up to be, a watched pot will never boil.

But the statement "a watched pot never boils" infers that an *unwatched* pot *will* boil. To prove this is much more difficult. It involves a sort of Pot Uncertainty Principle. As soon as one checks to see if the pot has boiled, the pot becomes a watched pot and can not, therefore, be boiling. The only way to prove that an unwatched pot boils would be to leave the pot on the stove unwatched, and then return later to find it gone. If this occurs, it can only be assumed that the pot has boiled away, for what else could account for its disappearance?

This second experiment was performed successfully shortly after the first was completed. Once again I filled a pot with water and set it on the stove, but this time I left the apartment to go shopping all morning. When I returned home I found the pot was gone. The stove was also off, much to my surprise. I later inquired of my roommate and found out that she had turned the stove off as soon as she realized that there was nothing really on it, and that if I kept trying to burn down the apartment like that she was going to make sure I got evicted.

So it seems logical to conclude that the pot boiled away and my roommate found the stove without anything on it before turning off the gas. However, the next evening I found what looked like the very same pot that had boiled, sitting up in the cupboard, minus the deposits left on the inside from all the water that had boiled in it. At first this seemed to cast doubt on the results of my experiment, but I soon realized that the pot

in the cupboard could very well be a virtual pot that had been forced into real existence by the departure of the first pot from this universe, in accordance with the conservation law that states that the amount of cupboard space you have is never more than exactly the amount of cupboard space you need. By boiling the pot I had put the kitchen out of balance, so a virtual pot condensed out of the vacuum to fill the extra cupboard space.

So it is true that a watched pot never boils, and there is some good evidence that unwatched pots do boil, even though the latter phenomenon can never be observed directly.

Dr. Iva P. Aitchdee has been on the faculty at ULAC for approximately 3 months. She is the current director of the Schiplee Center for Immaterial Research. She is not well known for her pioneering work in the new field of quantum relativistic thermo-space-time dynamics.

The Physiology of Profanity

J. L. Kersting, Permafrost, Minnesota

Introduction

In defense of cigarette smoking, a disenchanted wag once noted that there are so few remaining viscerally satisfying things which one can do in public: it is no longer socially acceptable to openly scratch one's privates, fart, or pick one's nose. Society has, however, been increasingly cognizant of the value of irreverence in coping with the ever-greater levels of individual and interpersonal stress associated with modern living. As such, from all observable indications, there seems to exist a continually greater tolerance for the general use of blasphemy, obscenity, and all manner of creative desecration.

A preliminary casual and random survey of individuals[1] has been conducted which posed the question: Why do you use foul language? Overwhelmingly, the responses could be paraphrased as *"Because it feels good"*. This phenomenon will be a well-known dynamic to all, and any sociologic trend of this magnitude cannot help but pique the spurious attention of the interested clinician. Here at the Center for Unceremonious Research into Sacrilegious Efficacy (CURSE), we have taken these queried individuals at their word and sought to investigate the apparently real and ameliorating physiologic effects of profanity.[2]

Method

A randomized legally-blind study was composed using 24 adult volunteer subjects.[3] Candidate selection was limited to those with a maximum intraocular distance of 4 cm, maximum distance of 3 cm from upper orbits to hairline, and whose lips moved when they read. Special preference was accorded in selecting those volunteers who absorbed an average of >35 hours of prime-time television per week.

The study design called for individual subjects to be seated in a dark room before a projection screen. The subjects were fitted with electrodes and monitors to measure blood pressure, heart rate, electro-encephalogram, and galvanic skin reflex. Sphincter tone electrodes (capacitance strips embedded in adhesive tape) were offered to, but declined by, most subjects. Physiologic data were fed into a multi-channel strip-chart recorder for hard-copy and trending.

A series of 30 numbered slides was presented to subjects, in which a frustrating event was taking place to an individual, for example, bottom of grocery bag dropping out and spilling contents, and the late discovery of absent toilet tissue in a public restroom. Subjects also received a handout listing a selection of numbered epithets correlating to the viewing of each numbered slide. For example, upon viewing situational slide #1, the handout presents 3 expressions keyed to the fictional event – perhaps:

1 (a) "Aw, Shucks"
1 (b) "Well, Shoot"
1 (c) "Oh, Shit"

After viewing each slide intently for 2 minutes, subjects' instructions called for them to emphatically voice each exclamation choice aloud, pausing for one minute between phrases. The physiologic data strip was carefully monitored to mark at which point on the recording paper each vocalization was made during the presentation of a given slide.

Complications of the study were few. One hirsute male sustained an acute vagal episode of asystole following removal of the sphincter tone tape, and required 3 direct-current countershocks of 650 watt seconds, as well as emergency ballpoint pen tracheotomy. Another subject turned into a pillar of salt following the reading of one of the more lascivious execrations. The subsequent wait for Housekeeping staff delayed the progress of the experiment.

Results

Without exception, the presentation of each discomfiting slide elicited physiologic distress in the subjects. Objective assessment of legitimate distress was considered to be 3 or more of the following:

- elevated blood pressure
- increased pulse rate
- loss of primary alpha EEG waves
- diaphoresis sufficient to lower galvanic skin resistance
- and, of course, sphincter tetany.

Analysis of the tabulated physiologic effects of the various vocal responses showed that, without exception, those physiologic indicators of distress were diminished within one minute of pronouncing the more scurrilous utterances within each subset of responses. In post-test debriefing, subjects confirmed the experimental evidence by unanimously volunteering that, subjectively, they "felt better" and were much more relaxed after the more sacrilegious remarks.

Discussion

The results of this study would seem to be too explicit and unequivocal to ignore. Clearly, further research will be necessary to elucidate to what extent this scatological phenomenon is either biological or psychodynamic.[4]

In consideration of this point, a brief excursion was made into the potential adaptiveness of what could possibly be an *innate* vulgarity. To this end, it was decided to take a look at those creatures with the least evolutionary distance from Man, the non-human primates. This seemed a potentially rewarding inquiry, as it is well known that psychologists and primatologists have been teaching non-verbal language skills to higher apes for some time now.[5] Consider these brief but striking preliminary findings:

- Rinsoe the chimp, raised by husband-and-wife psychologists in Nevada, has been taught American Sign Language since birth. She converses with her keepers, and has been celebrated for even devising vocabulary when necessary (upon seeing a duck for the first time, Rinsoe signed "water bird"). Recently it was discovered that Rinsoe has learned and invented some rather off-color words.[6] Her siblings and children have acquired Rinsoe's communicative signing ability by association, and have at times been caught sharing rude jokes.

- Kookoo, a signing gorilla raised since infancy by anthropologist Nancine Peterson, is shortly to have published his first volume of prurient limericks, *There Was a Young Langur from Madras*.[7]

- Lola, a chimp housed at the Xerxes Regional Primate Center in Atlanta, has been taught language capability by pressing sequences of illuminated buttons with various letters and syntax symbols printed upon them. In this fashion, Lola can converse, request food from a machine, and request the company of various trainers. It was only recently discovered that she has, in addition, assembled an impressive collection of lascivious insults.[8]

Notes

1. Manuscript in preparation. Backroom Poshe Lounge, Superior, Wisconsin.
2. Research supported by NIMH grant #B.S.-54321.
3. Members of the Minnesota Institute for the Preservation of Acephalic Ennui.
4. Funding bodies, please contact the author at Red Lace Massage, 5 East Superior Street, Duluth, Minnesota.
5. *Annals of Philologic Emulation* (APE), II, 3:35-9.
6. *Journal of Execration by Simians Under Siege,* VII, 2:109-18.
7. Hominid Press, 3508 Knuckle Walk Trail, Nairobi, Kenya.
8. Personal communications, D. Rickles, Beverly Hills, California, and J. Rivers, Hollywood, California.
9. Fundamentalist inchbrows please address churlish insults to C. Darwin, *Beagle,* Boat Well D-7, Galapagos.

Fun With Geometry

© James Stanfield, co-founder, The Institute for Further Research

The Perpendicularogram

Much has been made of the simple geometric figure, the parallelogram, where opposite sides are parallel – hence the name – and its cousins, the square and isosceles parallelogram or rhombus, where all of the sides have equal length.

I ask you, is perpendicularity such a poor second cousin of a geometrical relationship that it does not have a figure named for it? A search of the mathematical literature is resoundingly silent on the subject. Where is this missing figure, the perpendicularogram? Could it be that this simple figure with opposite sides being perpendicular does not exist? Poppycock! With ruler and compass in hand, here is how to construct one.

Beginning with a horizontal line of unit length, construct a second perpendicular line floating above it, and centered on the first line. Do not let these 2 lines touch! That would make them adjacent and not opposite sides.

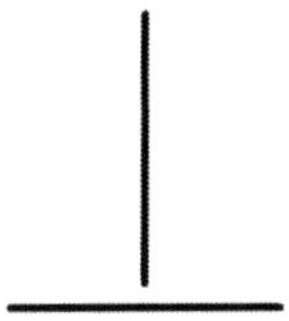

We are off to a good start, but what to do next eludes me. While I am waiting for inspiration, I will describe how to make an ellipse compass.

Figure 1

An Ellipse Compass

Compasses are widely used in geometry and many other fields to draw circles, which are just a special case of the ellipse in much the same way that the square is a special case of the rectangle. There is a tried and true way of drawing ellipses with a pencil, a loop of string, and 2 pins, but this method is cumbersome and imprecise. Wouldn't it just be a lot easier to draw one with some special kind of compass and be done with it? I have discovered a way! Indeed, it was my experience

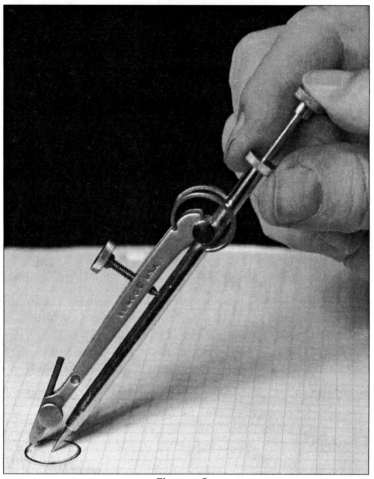

Figure 2

attempting to draw tiny circles with a regular compass
where the height of the lead was maladjusted to the
center point that I noticed a wobble whence the path of
the lead deviated from a perfect circular form, much in
the same way that the Sun wobbles into an elliptical
form during its journey round the Earth.

Conjecture here: could it be that the luminiferous
æther flows at a slight angle to the ecliptic, explaining
Albert Michelson's and Jacob Marley's null result? But I
digress.

For just the purpose of drawing tiny circles, the
draftsman prefers a drop bow compass where the lead is

held by an armature, which is free to slide vertically up and down the shaft of the center point, thereby precisely mating lead, center point, and paper.

The locus of all points of the lead as it freely rotates at a set distance from the shaft of the center point, and, at the same time, is free to rise and fall on that self same shaft, is a cylinder. If the shaft of the center point is held at an angle to the page, the intersection of the plane of the page with the cylinder is an ellipse.

Alas, this rise and fall is quite restricted, which limits the size of the ellipses achievable. Extending the length of the shaft would theoretically increase the size range of the ellipses that one could draw. Until I can afford to have a custom center-point shaft fabricated, I will just continue to use an ellipse template.

Construct a Golden Section, and From That, a Pentagon

Start with triangle ABC of Figure 3, where AB = 2x BC. Strike an arc CB from point C. The point where arc CB intersects line AC is point D. Strike an arc AD. The point where arc AD intersects AB is E which divides line AB into the *golden ratio*: EB/AE = AE/AB (the smaller

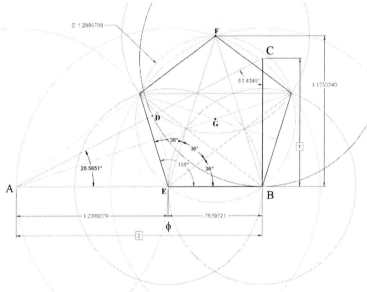

Figure 3

25

segment is to the larger segment as the larger segment is to the whole). Strike arc AE from E and B. Where these intersect at point F is the top of the pentagon with line EB as the base. Strike arc AB from E, B, and F. Where they intersect are the other 2 points of the pentagon. The perpendicular bisectors of lines BF and EF will intersect at the center of circle G, which circumscribes the pentagon.

A Method of Trisecting an Angle

Since trisection of the angle has supposedly been proven to be impossible, and the proof of this is beyond the scope of this article (not to mention that it is beyond the scope of the author's understanding), I will begrudgingly stipulate that it is, indeed, impossible. Therefore, my method must somehow lack the requisite rigor and perfection. But, dag-nab it, it's close enough for government work.

Starting with angle BAC to be trisected, construct circle 1, of convenient radius, R, from point A (see Figure 4). At points B and C, where circle 1 intersects the sides of angle A, construct 2 more circles, 2 and 3, also of radius R. Circles 2 and 3 intersect at points A and D. Construct line AD, which bisects angle A, and extend it to intersect the opposite side of circle 1 at E. Construct circle 4, of radius R, at point E. Extend line DE to

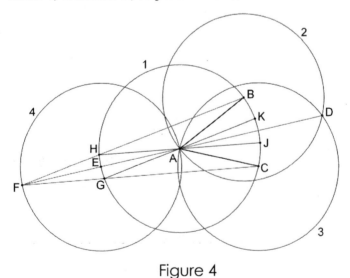

Figure 4

intersect circle 4 at F. Construct lines FB and FC. The points at which lines FB and FC intersect circle 1 are G and H. Construct lines GJ and HK through point A. Lines AJ and AK trisect angle A.

After that, figuring out how to complete the perpendicularogram should be a snap! In Figure 5 we begin by labeling the base line AB. Construct an arc of length AB from A which intersects the vertical line, which is our opposite side, at C. Construct the perpendicular bisector of line AC starting where it intersects the vertical line at D extending to where it seems to intersect AB at B! There must be some deep math here somewhere! All opposite sides are perpendicular. Lines AB and AC are equal as are lines BD and CD. Therefore, this is an isosceles perpendicularogram.

Here is some idle speculation. Since this figure is so simple, countless geometers, no doubt, would have discovered it; but the name is long and cumbersome. Indeed, I readily admit that the name "perpendicularogram" has no boogey. Therefore, I suggest that this figure be given a snappier and more memorable name. I suggest that it be called Bob.

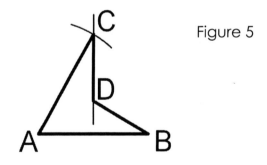

Figure 5

Niddy-Noddy-Noodle

Augustus De Morgan, University College, London

The great bulk of the illogical part of the educated community – whether majority or minority I know not; perhaps six of one and half-a-dozen of the other – have not power to make a distinction, cannot be made to take a distinction, and of course, never attempt to shake a distinction. With them all such things are evasions, subterfuges, come-offs, loopholes, etc. They would hang a man for horse-stealing under a statute against sheep-stealing; and would laugh at you if you quibbled about the distinction between a horse and a sheep.

I divide the illogical – I mean people who have not that amount of natural use of sound inference which is really not uncommon – into three classes: – First class, three varieties: the Niddy, the Noddy, and the Noodle. Second class, three varieties: the Niddy-Noddy, the Niddy-Noodle, and the Noddy-Noodle. Third class, undivided: The Niddy-Noddy-Noodle. No person has a right to be angry with me for more than one of these subdivisions.

– *A Budget of Paradoxes*, second edition, 1915, vol. II, p. 196.

Field Tests of the Efficacy of *Cakus chocolatus* for the Treatment of Disease

Jay M. Pasachoff, PhD, Naomi Pasachoff, and Eloise Pasachoff, Williams College

It is widely rumored that a large number of common diseases and ills respond favorably to a treatment of chocolate cake, *Cakus chocolatus,* taken internally once or twice a day, usually after meals. To test this hypothesis, a series of experiments was set up in our laboratory (converted from our dining room for the purpose).

Subjects were recruited randomly from the local population of college students. Every third passer-by was approached with the question "Would you be willing to participate in a laboratory experiment to test the efficacy of chocolate cake in curing whatever ails you?" It was found that 97% of those asked responded affirmatively to both this question and to the follow-up question "Do you have some minor ill that would make you eligible to participate in this experiment?" The other 3% all seemed torn, and paused before answering, but eventually muttered the word "diet", spun around, and ran off.

Subjects were divided into 3 groups. The first group consisted of 6 students and the 2 experimenters (who, as devoted to this study as any scientists, were willing to use themselves as guinea pigs in the name of Science). They were fed cubic pieces of chocolate cake, *Cakus chocolatus**, and were asked to consume it in a 3-minute period, along with their choice of coffee, tea, or milk. Each of the subjects was also supplied with a spoonful fo sugar (to help the medicine go down).

A second group of equal size, the control group, was seated at the table, and then given empty plates, forks, beverages, and spoonsful of sugar. As it was a single-blind experiment, they were not told whether they were the prime group or the control group, although some may have guessed from the difference in taste.

29

After a suitable wait, to allow the *Cakus chocolatus* to act, all the subjects were asked how it had affected them. All the subjects in group 1 felt that their general well-being had been aided by the application of chocolate cake, and all the subjects in group 2 felt decidedly worse off than they had been a short time before. We interpret this to mean that, on at least a short time-scale, the *Cakus chocolatus* has a beneficial effect.

The third group was used to test topical applications of the *Cakus* remedy. 3 subgroups were formed, totalling the size of the primary group. The chocolate cake was applied to the first subgroup in an axillary manner, with the measured cube of cake held under the left armpit for 60 seconds. None of these subjects reported the same beneficial effects that those who had taken the *Cakus* orally had reported.

Another subgroup had the *cakus* applied facially, in the manner illustrated so well in the work of Keaton (Buster Keaton, films, 1917 *et seq.*). They reported that all-in-all they preferred Boston cream pie.

Subjects in still another subgroup, who had the *Cakus chocolatus* applied anally, refused to answer any questions from the experimenters, and in fact have not spoken to us since.

As a result of these experiments, we can unhesitatingly recommend the oral application of *Cakus chocolatus* as a general remedy.

------

*Although the generic name is given in this article, we actually used *Cakus chocolatus* from Sara Lee Laboratories, which nobody didn't like. The relative efficacy of various competing brands of this powerful medicine is the subject of a current series of experiments that we are carrying out. As we are not restricting our experiments to the possible effects of these remedies on non-malignant diseases, we shall apply to the National Cancer Institute for grant support.

☛ What other substances could you experiment with?

The Franklin Twins

William DeBuvitz

Men and melons are hard to know.

Benjamin Franklin was truly a remarkable man. And, despite all of the writings by him and about him, there are still aspects of his life that are not completely clear. He had so many different talents and so many different professions it is hard to believe that one person could have accomplished all of this. How was it possible? Many writings about Franklin say he did the work of 2 men. And consider this: There is a University of Delaware website that tabulates in detail all of Franklin's activities when he was in Philadelphia.[1] Listed for 1744 are these 2 items:

(1) On October 25th or 26th Benjamin Franklin journeyed to New York City and returned on November 2nd.

(2) On October 26th Benjamin Franklin attended a Junto meeting in Philadelphia. (The Junto was an intellectual and benevolent society Franklin created in 1737 that had meetings on Friday nights.)

How could Franklin have attended a night meeting in Philadelphia while traveling to or having already arrived in New York City? Unless he had access to the Amtrak *Acela Express* train, he would have found this quite impossible. Historians probably dismissed the above contradiction as simply an error in the historical records of that time, but perhaps there is another explanation. To quote Sherlock Holmes, one of the great twins of detective fiction: "Eliminate all other possibilities and whatever remains, no matter how strange, must be the answer."[2] In this case, the answer is: <u>Benjamin Franklin was really 2 people – identical twins.</u>

There was the Left-Brain (LB) Franklin, the logical, objective, and analytical scientist. And, like many scientists, a bit shy. Then there was the outgoing, intuitive Right-Brain (RB) Franklin, the statesman and writer. Based on my studies, RB Franklin was the better-known of the twins. He was the one who did most

31

of the writing and lived the longest. LB Franklin died much earlier, as we shall see.

Left-Brain Franklin   Right-Brain Franklin

As everyone knows, any historical event or time period can be interpreted in various ways and if you look at Franklin's life with the right "attitude", the idea of twins becomes pretty obvious. RB Franklin enjoyed creating hoaxes – his publishing of an edict by the King of Prussia proclaiming that Britain was actually a colony of Prussia was one of his best. He also liked to use various aliases in his writings (Silence Dogood, Richard Saunders of *Poor Richard's Almanac*, Martha Careful, Celia Shortface, etc.), so hiding the fact that he was a twin would have come naturally. And with his sense of humor, he could drop hints about twins in his writings, including his proverbs or maxims. Also, I believe that if Franklin were alive today he would approve of this article and **JIR**'s many pseudonyms. He would probably be a major contributor to **JIR**, maybe even editor and publisher.

Also, it is worth noting that many of the proverbs sound like the work of someone with a dual personality – they refer to a single person as if he were 2 people. Is this a hint about twins?

How was a person so admired and respected as Franklin, able to hide the fact that he was a twin? RB Franklin might have answered this himself:

Admiration is the daughter of ignorance.

32

Perhaps another reason has to do with the limited communication that existed in the 1700s: 2 people in different locations might have each seen one of the Franklin twins and never had a chance to compare their observations. And, except for that one minor slipup on October 26, 1744, the Franklin twins seemed to do a good job coordinating their appearances so they were never both seen in public at the same time.

An important question to ask is, did his wife, Deborah Read Franklin, know of the twin? Probably. It appears that she was married to the longer-lived RB Franklin and went along with the secret. And perhaps she simply followed Franklin's maxim:

Keep your eyes wide open before marriage, half shut afterwards.

Let us now go through his (or their) life (or lives) and look at the evidence:

A. THE CONFUSING BIRTH AND CHILDHOOD

'Tis easier to build 2 chimneys than to keep one in fuel.

The historical records say that Benjamin Franklin was baptized almost immediately after he was born. The church happened to be located across the street from his home. The birth was on a Sunday, January 17, 1706. His father quickly covered him in thick blankets since it was winter and took him across the street to the church for the baptism. It was a very fast ritual and a bit confusing to those present. Why was it done so fast? Could there have been 2 babies in the blankets?

But why hide the fact they were twins? Perhaps it had to do with the house his parents were renting. When Benjamin Franklin was born, there were at least 12 people living in a small rented house. There were at least 10 children (the number varied as the older ones came and went), together with other relatives. The question is, how many people would the landlord allow to live in this house? Perhaps the addition of twins was one too many, so they announced the birth of only one son so they would not be evicted.

B. THE CONFUSION ABOUT HIS SCHOOLING AND MATHEMATICAL SKILLS

What maintains one vice would bring up 2 children.

The parents could barely afford to send one of them to school, let alone 2, but they tried. And their method was ingenious: They had the twins attend on alternate days! Since they were identical, the teacher never knew this. After each day of school, the twin that attended would "teach" the other that day's lesson. As you can see in the maxim below, it did not work out very well and their schooling ended at the age of 10:

He that teaches himself hath a fool as a master.

Notice the dual personality in this maxim.

Also, there seems to be a contradiction concerning Franklin's mathematical aptitude. He was poor in mathematics when he was in school, but showed great ability later in life. He said he taught himself. Perhaps this was more effective than alternating school days with his twin.

C. THE PRINTING BUSINESS

As is well known, Franklin moved from Boston to Philadelphia. This was in 1723, when he was 17 years old. His twin could have entered the city later and from a different direction. When he opened his printing shop in Philadelphia, people noticed that he was working continuously from day to night. One of the town's prominent merchants was quoted as saying: "The industry of that Franklin is superior to anything I ever saw of the kind; I see him still at work when I go home from club, and he is at work again before his neighbors are out of bed."[3] People said he was doing the work of 2 men, which makes sense since "he" really *was* 2 men.

Perhaps things did not work out so smoothly later in the business as implied in the following maxims:

The rotten apple spoils his companion.

Be ashamed to catch yourself idle.

I'll bet a lot of **JIR** readers would use these maxims to describe a brother or sister.

D. THE FAMOUS LIGHTNING EXPERIMENTS – THE DEATH OF A TWIN?

Wise men learn by others' harms, fools scarcely by their own.

According to historical records, Franklin was involved in a lot of scientific research leading up to the lightning experiments. After that, his scientific work decreased and there was a good reason: The scientific LB Franklin was killed during the first lightning experiment. The evidence is pretty strong:

Franklin's early electrical experiments were a bit reckless, but no one at that time knew the dangers of electricity. He used a primitive form of capacitor called a Leyden Jar (a glass jar with metal foil on the inside and outside to store charges) to do various experiments, some destructive. He tried killing chickens and turkeys with electricity and succeeded with the chickens, but had trouble with the turkeys: "But the turkeys, though thrown into violent convulsions, and then lying as dead for some minutes, would recover in less than a quarter of an hour."[4]

He then linked his electrical jars together, called it a battery and found it more "effective": "We killed a turkey with them of about 10 lb. wt. and suppose they would have killed a much larger. I conceit that the birds killed in this manner eat uncommonly tender."[5]

And his experiments went beyond just fowl: "I found that a man can without great detriment bear a much greater electrical shock than I imagined. For I inadvertently took the stroke of 2 of those jars through my arms and body, when they were very fully charged. It seemed a universal blow from head to foot throughout the body, and was followed by a violent quick trembling in the trunk, which gradually wore off in a few seconds. .... My arms and back of my neck felt somewhat numb the remainder of the evening, and my breastbone was sore for a week after, as if it had been bruised. What the consequence would be, if such a shock were taken through the head, I know not."[6]

35

I believe he found the answer to that last sentence a
few years later. Franklin actually developed *2* lightning
experiments to demonstrate that lightning was a form of
electricity. Here is his description of the first one, which
became known as the *Philadelphia Experiment*:

"On the top of some high tower or steeple, place a kind of sentry box big enough to contain a man and an electrical stand. From the middle of the stand let an iron rod rise, and pass bending out of the door, and then upright 20 or 30 feet, pointed very sharp at the end. If the electrical stand be kept clean and dry, a man standing on it when such clouds are passing low, might be electrified, and afford sparks, the rod drawing fire to him from the cloud. If any danger to the man be apprehended *(tho' I think there would be none)* let him stand on the floor of his box, and now and then bring near to the rod, the loop of a wire, that has one end fastened to the leads; he holding it by a wax-handle. So the sparks, if the rod is electrified, will strike from the rod to the wire and not affect him."[7] (I italicized the phrase in the above quote.) This was truly a very dangerous experiment. If ever there was a need for the warning, DON'T TRY THIS AT HOME, the above description had to be it, even with the addition of a wax handle.

A number of people in Europe tried the Philadelphia Experiment and got the results that Franklin predicted, but at least one died when he was struck by lightning. Oddly enough, Franklin never said he did this experiment. But it is hard to believe that a man with the scientific curiosity of Franklin would not have tried it himself. It seems much more likely that the scientific LB Franklin *did* try it — and died in the process.

His second "improved" experiment involved flying a kite in a lightning storm. He reported on it after news of the success of the Philadelphia Experiment in Europe had reached America. He described it in an odd manner, in an impersonal, third-person form, as if someone else performed it. Perhaps someone else *did* perform it. After seeing his twin, LB Franklin, killed by lightning, it seems reasonable that RB Franklin would be very reluctant to try any experiment involving lightning. Here is how Franklin began his description of this second lightning experiment:

"As frequent mention is made in the news papers from Europe, of the success of the Philadelphia Experiment for drawing the electric fire from clouds by means of pointed rods of iron erected on high buildings, etc. it may be agreeable to the curious to be informed that the same experiment has succeeded in

Philadelphia, tho' made in a different and more easy manner, which one may try, as follows."[8]

He then went on to describe the kite experiment as if someone else had done it.

After these experiments, his scientific work decreased and his statesman work really took off. This should be expected if the scientist LB Franklin had been killed and the statesman RB Franklin went on to pursue his own interests.

E. HOW DID THE TWINS' SECRET LAST FOR SO LONG?

Perhaps the following maxim might have "encouraged" people to keep the secret:

3 may keep a secret if 2 of them are dead.

RB Franklin certainly had a way with words, didn't he?

Well, now that you know that Benjamin Franklin was a twin you might want to take a second look at some other famous people. I will be too busy for this because I will be contacting my long-lost twin after he finishes his next James Bond film.

References

1. www.english.udel.edu/lemay/franklin/1744.html Web pages 1 and 23.
2. This statement and variations of it were made by the first Sherlock Holmes twin in *The Sign of Four* and *The Beryl Coronet.* They were repeated by the second Holmes twin in *The Blanched Soldier* and *The Bruce-Partington Plans.* It should be obvious to anyone who has read all of the stories of Sherlock Holmes that he truly died at the end of *The Final Problem* when he fought Professor Moriarty at Reichenbach Falls. In the stories that came later, there are subtle differences in the Holmes described by Sir Arthur Conan Doyle, which some people attribute to experiences Holmes had during the years he was missing. It seems more likely that this "changed Holmes" was simply another person, a twin brother who did not live in London until after the death of his twin. I will leave it to others to make a more detailed study of the evidence for the Holmes twins.
3. Walter Isaacson, *Benjamin Franklin – An American Life* (Simon & Schuster, 2003), p. 54.
4, 5, 6. H.W. Brands, *The First American – TheLlife and Times of Benjamin Franklin* (Doubleday, 2000), p. 198.

7. Isaacson, p. 138.

8. Thomas Fleming, ed., *Benjamin Franklin – A Biography In His Own Words* Volume I (Newsweek Book Division, 1972), p. 92.

Photos: Creative Commons Attribution 3.0 Unported. Sculpture by James Peniston, Girard Fountain Park, Philadelphia.

IMPROVISED MENTOS/SODA BOMB

Howard L. McPherson, Moab, Utah

Experts said a suicide bomber exploded a truck containing an estimated 400 kg of Mentos candy mints and 4,000 gallons of cold, CO_2-pressurized soda pop. Besides the bomber, 3 civilians were killed. A further 270 people were injured. DoD spokesofficer Howard McPherson said this brings a new level of terror. "The use of candy/soda in this exceedingly vicious viscous attack is deplorable."

For several days after the explosion, tens of millions of terrorist army ants were seen converging on the area.

A team of scientists has been sent to the bomb site to collect samples of the sticky residue. A full panoply of methods will be used to analyze the mess, including IR, UV, NMR, HPLC, AA, ICP, NAA, STM, CD, ORD, and MS. It may take a year to determine the flavors of the candy and soda. It is not yet known if the soda had been laced with toxic materials.

People trying to buy more than 3 rolls of Mentos candies (any flavor) will now be required to undergo a background check. The President has directed the ATF agency to change initials to ACTF to reflect its new roll in candy regulation. The DoD has set up a group to search for terrorist candy stockpiles.

From deep inside a cave, behind a large spider web veil, Osama Bin Laden released another video tape. "We have been thwarted in trying to kill the infidel with lead toys. Now we are going after their sweet tooth." Experts have said it may take a year of analysis to determine what this statement really means.

A team of diabetes counselors has been sent to comfort the victims.

THE CROSSWORD PUZZLE FROM HELL

Karl M. Petruso, PhD, University of Texas at Arlington

ACROSS

1 The _____
5 Tributary of the Bug (var.)
10 See 34 Down
14 See 10 Across
15 Opposite of "average"
16 Hebrew month
17 Part 2 of quote

DOWN

1 Opposite of "forty"
2 Common Burkina Faso insect
3 Part 3 of quote
4 Atlantis to Xanadu dir.
5 A 5-letter word
6 "... of _____"
7 Hebrew month

19 An adjective
20 Madagascar roofing tool
21 See 23 Down
22 Color I am thinking of
24 Color I will be thinking of
   for tomorrow's puzzle
26 John _____
27 See 59 Across
29 A noun (misspelled)
30 Arlington to Springfield dir.
33 Crater on Mars (abbrev.)
34 "... a _____" (Keats)
35 Article abroad
36 For _____
37 "... begat __" (Book of Genesis)
38 Yiddish for "putz"
39 See 12 Down

40 ____ rrështl (Albanian
   board game)
41 It's bigger than a molecule
42 Airport abbreviation
43 Popular Lithuanian female
   nickname
44 Ancient Bactrian bronze coin
   denomination
45 2 + 2, e.g.
47 With 5 Down, car part
48 Speaker of quote
50 Year Lucius Mummius
   Pulcher almost remarried
51 Mars to Neptune dir.
54 Agua, in Acapulco
55 Beginning of quote
58 Bodkin (archaic)
59 See 27 Across
60 Random string of four letters (var.)
61 He batted .219 in 1953
62 A color I am not thinking of
63 _____ Jones

8 "... and _____"
9 Person who did not speak quote
10 Cousin in Ulan Bator
11 Appendix to quote
12 See 39 Across
13 Random string of four letters
18 King _____ II
23 See 44 Down
25 Synonym for 19 Across
26 Antonym for 25 Down
27 In _____
28 Medieval Uzbek horse bit
29 Hebrew month
31 Theme of this puzzle
32 Theme of a different puzzle
34 See 10 Across
37 Another person who did not
   speak quote
38 Assistant key grip, *Forrest
   Gump*
40 Hebrew month
41 What 26 Across was famous for
44 See 21 Across
46 "... good ____" (Shakespeare)
47 "I'm gonna sock you in the
   _____, Will Shortz."
48 Prefix suffix
49 "Uh-oh" (colloq.)
50 Hebrew month (var.)
52 Poetic contraction
53 Poetic expansion
56 LP:CD::DVD: _____
57 Burma to Myanmar dir.

A Rebuttal to Multiplication

James Cargal

Introduction

It utterly dismays me to see a work like Son-of-Son-of-Ag's "Multiplication: A New Operation"[1] on a cave wall. When I see an article such as that, I know we must be entering a new Ice Age. While it may have an insect's-spit-worth of abstract interest, it is without any other redeeming value, as I shall show.

I. Originality

The "new" operation is not original. It is merely a speeded up addition. For example, $3 \cdot 4$ is just $3+3+3+3$ (4 times).

II. Usefulness

Whereas $3+3+3+3$ may be handled by the "new" operation, $3+4+5$ cannot. All the numbers being added have to be the same!!

III. Practicality

i) To multiply, one must learn an entirely new composition table for $X \cdot Y$. And, the only case where $X+Y = X \cdot Y$ is when $X = Y = 2$ $(2+2 = 4)$.[2]

ii) One has to learn how to mix $\cdot$ and $+$. For example, $X \cdot (Y+Z)$; $X+(Y \cdot Z)$; $X+\cdot Y$; $X \cdot +Y$; and more! Son-of-Son-of-Ag solved *none* of these problems.

iii) The multiplications table involves huge numbers. For example, $6 \cdot 7 > 10+10+10$.

IV. Theory

i) Son-of-Son-of-Ag asserts without proof that $X \cdot Y = Y \cdot X$. This would mean that for any X and Y, $X+X+X+...+X$ (Y times) $= Y+Y+Y+...+Y$ (X times). Counterexamples are so numerous that it is not worthwhile to bother with them.

ii) We can show that Son-of-Son-of-Ag's idea is ridiculous, *reductio ad absurdum,* by extending the method. We let $X^Y = X \cdot X \cdot X \cdot ... \cdot X$ (Y times). What use is an operation like X^Y except to give us absurdly large numbers?

V. Premise

Whereas Son-of-Son-of-Ag's multiplication cannot even solve a problem like 3+4+5 once, Another-Son-of-Big-Thunder-the-Greatest-Stud-in-the-Universe-Past-Present-Future-and-Otherwise-Not-to-Mention-Being-Able-to-Crush-the-Thighbone-of-a-Bison-with-his-Jaws has shown that 3+4+5 can be solved in two other ways:

a) as 4+3+5;

b) $3+(4+5)^3$ (this second is only understood by three or four Cro-Magnon alive).

VI. Conclusion

Clearly, the caves should devote their space to work like Son-of-Et-Cetera's rather than spurious nonsense such as Son-of-Son-of-Ag's multiplication.

References

1. Ag, Son-of-Son-of. *Multiplication: A New Operation.* Peoples Democratic Cave, Year of the Two-Headed Wolf.
2. Pig, Pig, *2+2 = 4, Probably*, Four Caves, Rechiseled at Smelly Creek, Year of the Hot Winter.
3. Being-Able-to-Crush-the-Thighbone-of-a-Bison-with-his-Jaws, Another-Son-of-Big-Thunder-the-Greatest-Stud-in-the-Universe-Past-Present-Future-and-Otherwise-Not-to-Mention, *Introduction to the Elements of the First Principles of 3+4+5*, Cave of White Bats, Year after the Year after the Year of the Hot Winter.

☞ Can you rebut Division? Calculus?

Inorganic Elemental Chemistry

G. Dinsburg

Elemental Chemistry is a new subject. This paper, obviously a milestone in the progress of Science, serves as the first disclosure of this important new field. Elemental Chemistry is not synonymous with Elementary Chemistry. This is not to say that it is a difficult field. Anyone with an IQ of about 150 can readily grasp its implications. Nor is Elemental Chemistry synonymous with Nuclear Chemistry although in this paper we also accept the Periodic Table of the Elements. Nuclear Chemistry requires energy sources which are getting more and more difficult to come by if one wants to work on the frontiers of Science. Elemental Chemistry is not hampered by the need to spend megadollars on buildings and equipment. Anyone with clean paper and a word processor, typewriter, or indeed paper and pencil, a suitable IQ*, and reasonable originality and imagination, is in a position to make important contributions to this new field.

Elemental chemical reactions of all types are very simple to execute. For the sake of simplicity, it behooves us to deal with them under various classes. We shall not attempt to cover all the known reactions.** Specific examples will suffice to show the principles involved.

I. Decomposition Reactions

1	As	= A + S
2	Ba	= B + A
3	Be	= B + E
4	Bk	= B + K

5	Ca = C + A
6	Ce = C + E
7	Cf = C + F
8	Cu = C + U
9	Eu = E + U
10	Fe = F + E
11	Hf = H + F

These examples illustrate both the theoretical interest and the technological applications of elemental chemical reactions.

It should first be noted that although in principle these reactions obey the elementary laws of chemical equilibrium, in practice it is much easier to carry them from left to right, as written, rather than from right to left. This is due to the formation of a gaseous element at standard conditions. Thus, high pressure would be required for many of the reverse reactions, unjustifiably increasing the equipment costs.

Reactions 8 and 9 make it possible to prepare uranium from readily available starting materials (particularly reaction 8). Reaction 9 enables one to obtain enough einsteinium to satisfy all those who wish to investigate its properties more fully, but the alternative, 8, particularly if one carries it out in conjunction with 11, makes it possible to prepare uranium hexafluoride in a hydrocarbon atmosphere. Thus:

$$Cu = C + U$$
$$\underline{6Hf = 6H + 6F}$$
$$Cu + 6Hf = UF_6 + CH_4 + H_2$$

This sequence also shows how elemental chemical equations are balanced. It is not worthwhile to carry out reactions 7 and 8 concurrently, as californium is hard to obtain.

When reaction 11 is carried out in the absence of copper, a side reaction, also known from Elementary Chemistry, occurs, obviously because hydrogen and fluorine atoms are involved:

$$Hf = H + F = HF$$

However, in the presence of 1 gram-atom of copper, 6 gram-atoms of hafnium react as shown above.

II. Double Decomposition Reactions

Discerning readers noted that the decomposition reactions were listed in alphabetical order. This appears to be the fairest way. Yet some license will be taken with double decomposition reactions for reasons which will become apparent.

Perhaps the most interesting reaction in this class involves aluminum and copper:

12          $Al + Cu = Au + Cl$

and not exclusively for mercenary reasons! (see acknowledgement) Here again, the formation of gas (chlorine) causes the equilibrium to be displaced as written, from left to right. It is improbable that this equilibrium will be studied in detail. In the first place, history has shown that there is greater interest in gold than in aluminum or copper. It is further of historic interest that the great alchemists searching for the mythical "philosophers' stone" did not grasp (these pioneers preceded Mendeleev) how close they might have come to realize reaction 12, Al being part of their name!

Certain professors of inorganic chemistry who love to regale their students with lecture demonstrations would note the following suggestions:

Decompose sodium under a bell jar with an opening on top fitted with a one-hole rubber stopper, through which is placed a piece of glass tubing connected to a piece of rubber tubing kept closed by a pinchcock.

When the sodium has decomposed, as evidenced by its disappearance (that is all you can see because an equimolar mixture of nitrogen and argon is colorless), open the pinchcock and introduce an equivalent amount of chlorine into the bell-jar. White flakes form and give the students the impression of a snowfall. This demonstration is reminiscent of toy "snowfall paperweights". That is safer to carry out than the alternative classical inorganic method of dropping of a sodium pellet into an atmosphere of chlorine. The reaction involved is:

12a $N + A + Cl = NaCl$

Although this reaction can be carried out at atmospheric pressure for the purpose of that demonstration, here again the equilibrium lies rather from right to left:

12b   NaCl = N + A + Cl

It must be admitted that some thermodynamicists have taken issue with this statement, citing the great stability of NaCl. They are however quite wrong in trying to apply the tenets of classical thermodynamics to elemental inorganic chemistry. The first law of elemental thermodynamics states "It ain't necessarily so."

---

*   Below 150
**   Details of 2 important reactions are still classified: $Np = N + P$; $Pu = P + U$. These reactions also have biochemical significance.

Acknowledgement

An ethical scientist always gives credit where credit is due. Credit here is due to G. Dinsburg. However, since the author maintains high ethical standards, he must give credit to an anonymous student who was barely bright enough to write reaction 12 on the blackboard. It would have been too much to expect ordinary chemistry, too!

☛ **Suggest other interesting reactions to www.jir.com.**

IRREPRODUCIBLE MOLECULES

Alexander Kohn et al.

$CoFe_2$

Coffee

BaNa₂

Banana

H I J

K L M

N O

H_2O

Acetic acid

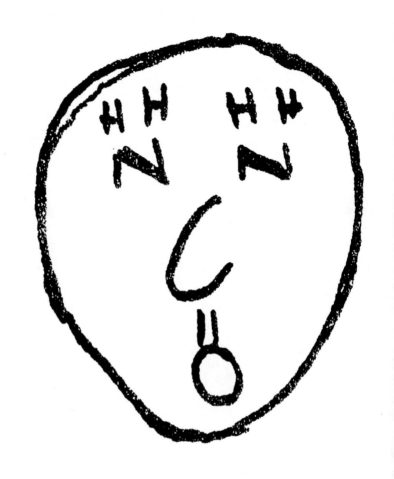

Urea

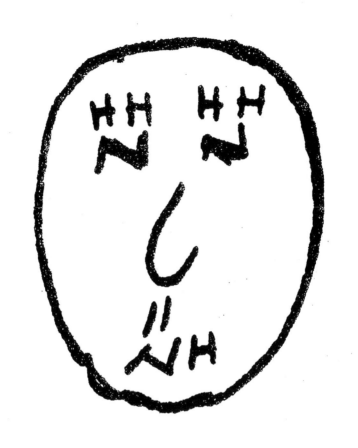

Guanidine

Cat-ions

Li+ ion

Counter-ion

Twitter-ion

Un-ion-ized

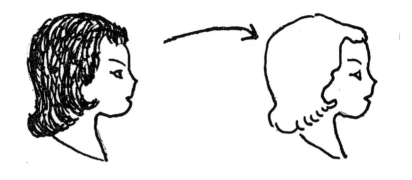

Oxidation

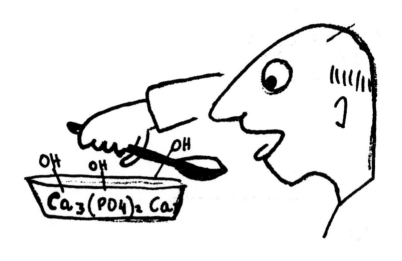

Hydroxy-apatite

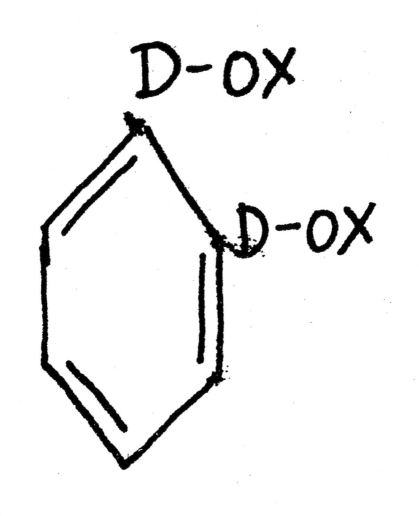

D-OX

D-OX

Orthodox

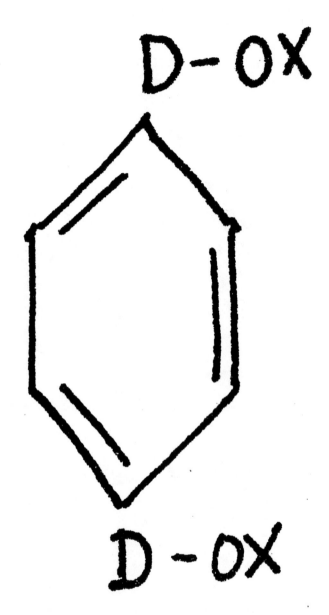

Paradox

Gold plated chain: Dr. Grubb

Di-hydroxy-chickenwire

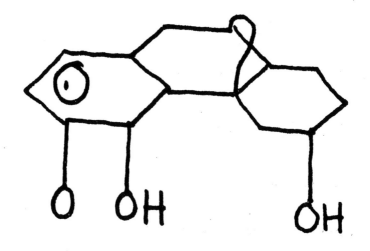

Morphine

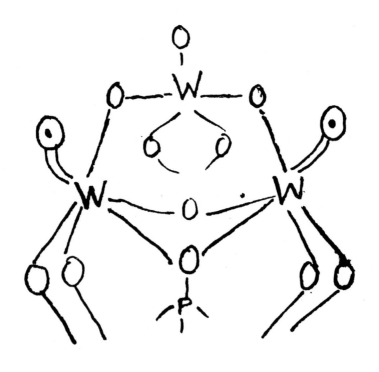

Phosphotungstate radical: Dr. Pease

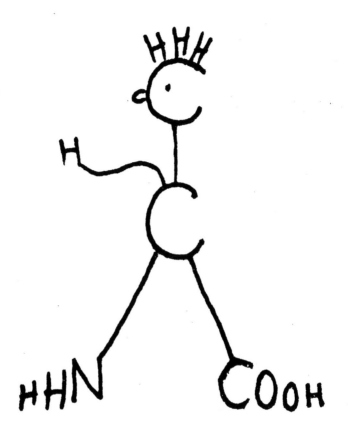

Alanine

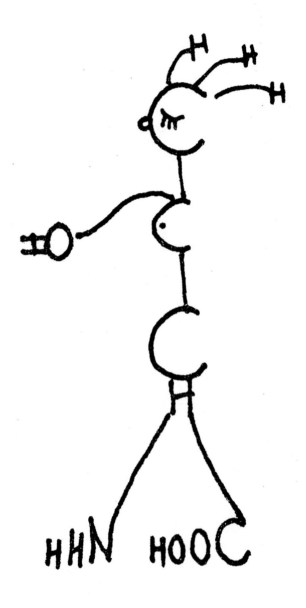

Threonine

Isoleucine

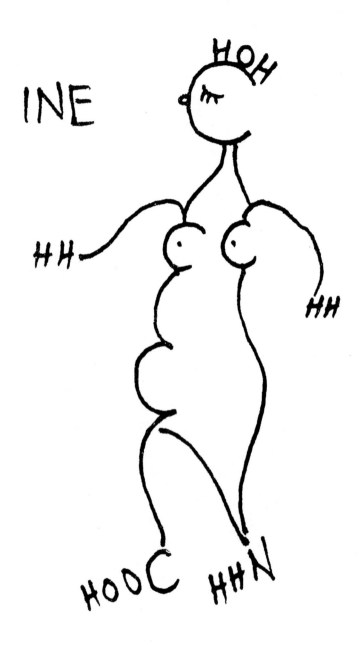

Hydroxyproline

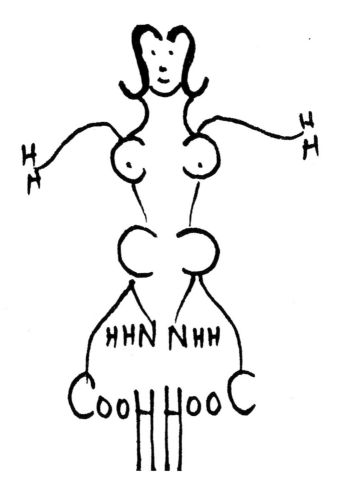

Cystine

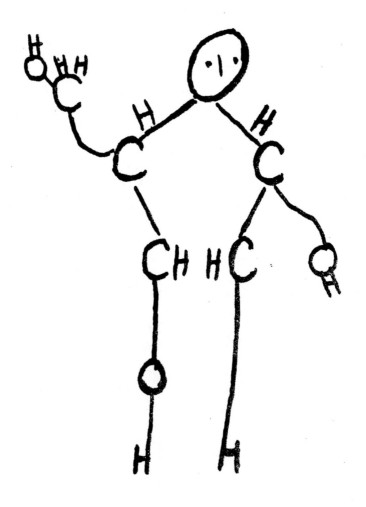

Deoxyribose

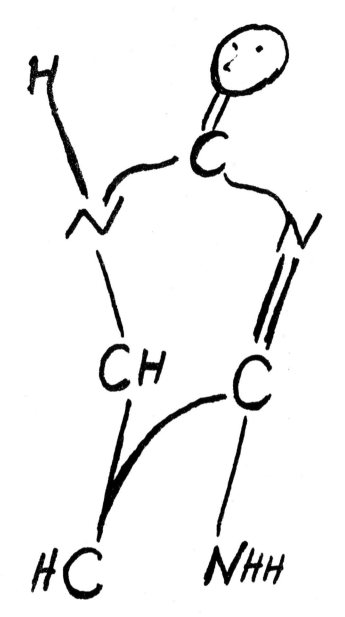

Cytosine

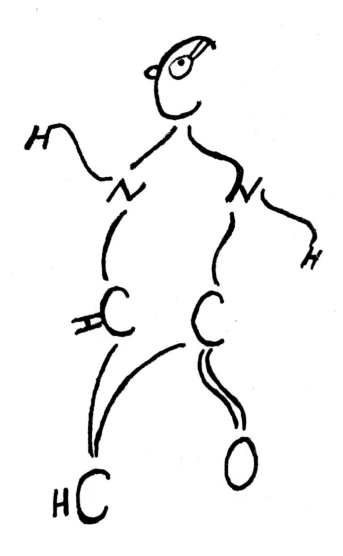

Uracil

Guanine

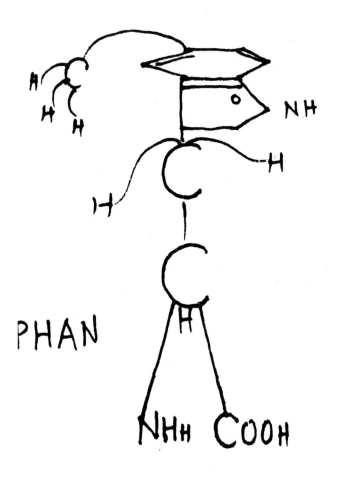

PHAN

Methyl Tryptophan

Dipicolinic Acid: Dr. Barzilay

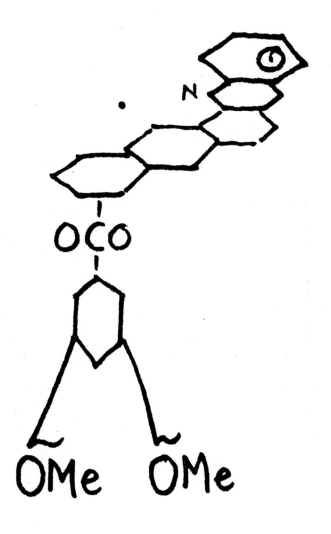

Reserpine

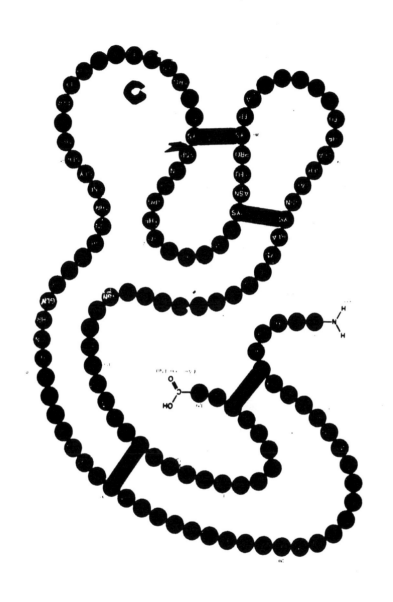

Lysozyme: Dr. Philips

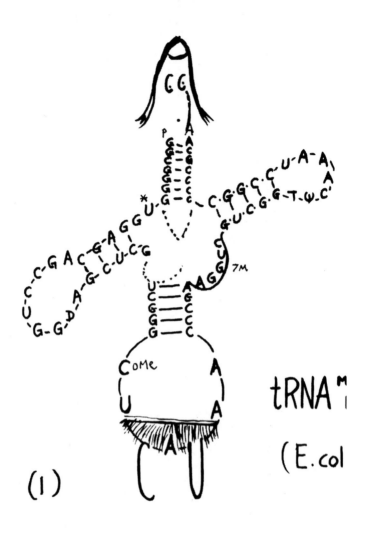

tRNA$^M_1$
(E.col

(1)

tRNA E coli

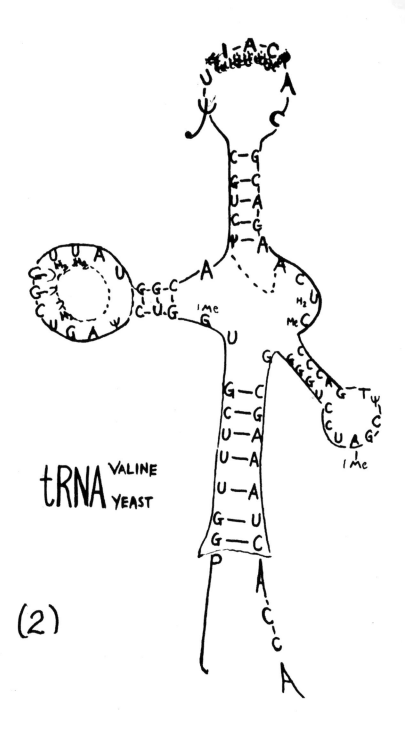

tRNA VALINE YEAST

(2)

(3)

tRNA SERINE (RAT)

Risk Factor Correlations With Respect To Cat Defenestration

Burton Deedrick Schmuck (Yes, that's my real name!)

Abstract

A proper ratiocination of cat defenestration, resulting in an empirically derived relationship allowing for the calculation of a unit-less risk factor for said defenestration, and correlation of that risk factor to an LD_{50} and other injuries.

Introduction

With the advance of the "information age" there has been a significant increase in the number of cat defenestrations[1]. In direct proportion, cat-inflicted injuries and fatalities have also risen. A need to predict the risk associated with throwing a cat out the window has driven this research project. Due to the serious need to prevent further *cata*strophic injury (where else do you think that word came from?) we have empirically derived the following formula (Equation 1) to predict the risk of defenestrating a cat.

$$R = \left(\frac{\alpha}{M}\right)^{\frac{B}{.112358}} - \left(\frac{T_{ch}^{-B}}{\Omega}\right) \qquad \text{Equation 1}$$

R – Risk Factor
M – Moxie of defenestrator
T_{ch} – Cat's developed trust for human defenestrator
α – Adroitness of cat
Ω – Cat's inherent supernatural instinct for danger
B – Bloodlust of cat (related to genetics)

Determination of α

There are perhaps thousands of methods for the determination of the natural adroitness of any given

---

[1] Some website on the Internet. Even though I can't find it again, it must be true because someone took the time to write it.

feline, however, Wassup et al.[2] standardized the process, based on observations from the annual Latah County (Idaho) cat toss. In summary, the process involves surprising a cat by launching it into a standard grove of pine trees[3]. The skillful use of agility under pressure (adroitness) can be measured as the inverse relationship of the number of branches the cat misses (or rebounds off of) before it is able to catch a secure hold.

The Bloodlust factor (B)

It can be commonly observed in even the most amiable of cats, that there are certain times when there seems to be a calculated decision that blood needs to be let from another entity. This has been shown to be due to the aristocratic nature of the species[4]. The more pure a cat's bloodline, the more "noble" it is. To show the power inherent in nobility, a cat will lacerate another entity, usually the one it considers to be its lowest servant (e.g. who buys its food). As a cat progresses "downward" in genetic refinement (for example: from Siamese to a common household pet), it has a less frequent need to mete out injury. However, since cats view themselves as the crowning glory of all creation[5], there will always be some need for this in every cat.

Bloodlust can be calculated as the reciprocal of the number of heterozygous allele pairs (H) divided by the total number of gene pairs (n) present in the cat (Equation 2).

$$B = \left(\frac{H}{n} \right)^{-1} \qquad \text{Equation 2}$$

Moxie (M)

A simple definition of moxie is "guts" in the sense of bravery. It has been observed that displays of moxie are

[2] Wassup, Hey-Dude, and Lagerman, "Fun at Backwater Festivals." *Journal of Rural America,* vol. 52:07.1234, Kansas, 1999.

[3] Green, Arbolis, "A New Standard for Idaheathen Vegetation." *Journal of Research:* National Institute of Standards and Technology, Vol. 106 No. 4, 2001.

[4] Personal experience.

[5] Schmuck, Burton "Conversations With My Cat" Penultimate Press, Seattle, 1996.

directly related to gut size, with a certain lag time. In the military, it is common practice to promote a man after displays of bravery and daring exploits. It has also been known that the size of a man's paunch often increases with his rank[6]. Therefore, it can be concluded that a man's guts is directly related to his gut size. Moxie is placed in the equation due to the fact that a larger intestinal tract increases the cat's difficulty in eviscerating a man in totality (not to mention that it takes a lot of guts to toss Aunt Nora's cat out the window[7]). It can be determined by representing a person's belt size in the appropriate units.

The developed trust of a cat towards a human (T_{ch})

This factor can easily be determined with the distance a dozing cat will allow a "trusted" individual to approach with a bucket of ice water. The full details are explained in the literature[8]. It should be noted that an ice-water-to-cat distance of zero is considered to be an outlier. A repeat of the experiment will not give zero again unless the cat is dead instead of sleeping.

A cat's supernatural instinct for danger (Ω)

A correlation to predict this factor was developed from inspiration derived from Schrödinger's "cat in a box" experiment. It is undergoing peer review and we hope it will be published soon. The basic procedure involves measuring many physical properties of a cat (entropy, heat capacity, octanol-water partition

---

[6] A military source. Will provide reference upon verified Top Secret Polygraph security clearance.

[7] Trust me.

[8] Pooler and Cashell. "I Feel a Cold Rain Coming On" Washington State University, 2000.

coefficient, thermal and electrical conductivity, tensile strength, auto-ignition point, etc. ...), and then placing the cat in a thin opaque box not quite twice the length of the cat.

The box is then shot in a random location with a high caliber rifle, such that the chance (according to surface area) of hitting the cat is 1 out of 2. It should be noted that the ammunition used should travel at a velocity greater than the speed of sound to avoid providing any natural warning to the cat. Furthermore, star-shape hollow-point tipped bullets are useful so that there is no injury that might go unnoticed and cause error in statistical data collection[9]. It should also be obvious that the box then needs to be opened to determine the status of the cat (frightened or not).

It was found that all cats have some supernatural ability to avoid danger, as the number of injured cats per bullet was less than 50% in all cases. There was even one cat that avoided 495 shots. It is interesting to note that the 496th shot was a perfect number. There was also an inordinately high number of cats that did not dodge either the 6th or 28th bullet (also perfect numbers).

Results and Conclusions

In conclusion[10] it has been determined that it is indeed a risky business to throw a cat out of the window. Empirical observations show that reasonable risk factors range from 0.76 to 7892.3. In 50% of the cases, the cat broke skin at a risk factor of 1.2. Stitches were necessary at R-values of 2.54. Major organ damage occurred at 4.23, and death at 7.86.

Approximately 250,000 cats were defenestrated in the course of this research[11] and human subjects were criminals who volunteered to be part of a program investigating the role of animals in rehabilitation.

Special thanks go to Dr. Clifton M. Carey for his role in encouraging irreproducible results

---

[9] Peterson, Peterson, and Peterson. "How to Shoot a Cat" *Grandma's Farm Publications,* Cheeseland, 1987.

[10] Written this way to annoy every English professor in the world.

[11] Aunt Nora's was the most rewarding.

A Taxonomic Scale of Cluedness in Humans[1]

Robert E. Johnson

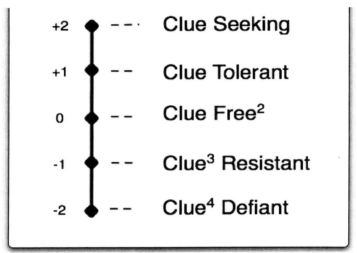

+2	Clue Seeking
+1	Clue Tolerant
0	Clue Free[2]
-1	Clue[3] Resistant
-2	Clue[4] Defiant

1 This scale attempts to correct the notion that cluelessness is a monovalent concept, lacking in a contrastive category. Until now, there was no antonym for "clueless", which may account for the smallness of the number of members of any committee who have one. "Cluedness" (pronounced klew-ed-ness) derives from the transitive verb "to clue-in". Thus, the small proportion of people who are clued-in are now identifiable. It also recognizes the fact that certain people are able to have a clue but sometimes don't.

2 Cluelessness can be taken to include all 3 of the lower forms in this taxonomy. It is clear that there are degrees of cluelessness, and that simply lacking a clue ("clue-free") is less common than is the aggressive refusal to accept a clue.

3 A "clue" is some piece of knowledge that may assist in the solving of a problem or some useful and accurate idea about how some system functions. A secondary meaning, "to have a clue", is to have a piece of

knowledge, as in: Q: "When did Bill get home last night?"
A: "I don't have a clue." Note that this does not define
the responder as "clueless". But to answer "3 in the
morning. He doesn't have a clue" does define Bill as
"clueless". Thus, if the subject of "to have a clue" is in
the first person, it is an example of the secondary
meaning, while if it is in the second or third person, it is
an example of the primary meaning.

4 All people who have viewed this scale have assumed
that they are +2, and have mentioned others who they
feel are –2. This was reciprocally so for all pairs of
subjects. "Cluelessness" may not be applied subjectively
as freely as it is applied objectively.

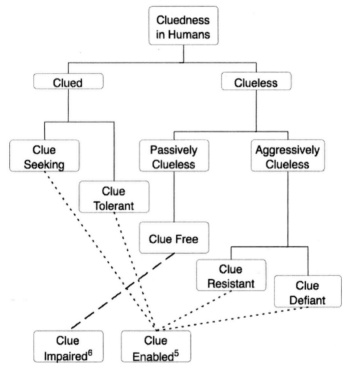

5 Note from the lower taxonomy that cluelessness is not
simply an issue of ability (Thanks to Flavia Fleischer for
pointing this out.) Certain humans appear to be able to
have clues, but either resist them or defy them. It is
likely that aggressive cluelessness is related to

complacency, expedience, or commitment to paradigm, each of which often leads to the denial of obvious clues.

6 It is politically correct to use either "clue impaired" or "clue challenged" in public discourse concerning a clue-free person. Referring to such a person as a "clue motel" is certainly impolite, and may border on being a hate crime.

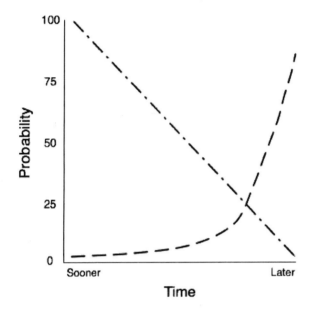

Legend:

— · — · — Likelihood of finding a clue that will be useful in solving a serious problem.

— — — — Probability of politicians or high-level managers seeking such a clue.

Mushroom Hunting for Beginners

© James Stanfield, co-founder, The Institute for Further Research

One of the fastest growing pastimes these days is mushroom hunting. And no wonder, what with the latest scientific findings that many mushrooms contain compounds that cure everything from cancer to the common cold. This, on top of the fact that mushrooms provide a delicious adjunct to any diet, explains why mushroom hunters are hitting the woods in record numbers.

Mycology (pronounced "mycology") is the technical term for the study of fungi. And study we must, for there is always the danger of mistaking the safe and delicious species for the poisonous and delicious species.

You wouldn't want to eat a toadstool, would you?

Most mushrooms have the following features in common: a disk-shaped cap sits on a tubular stalk; on the under side of the cap is a radial arrangement of gills; the gills produce the spores by which the mushroom reproduces. Some mycologists take the presence of these gills as proof that the mushroom evolved from the sea. But I digress.

Mushrooms may be hunted from January to December, but if you really want to find some you will want to hunt them during their growing season, which runs from the second through the fifth day following the first rain after the third Thursday in March.

The importance of

correct identification can not be overemphasized. It can mean the difference between basking in the afterglow of a fine meal of some of the finest delicacies that nature can provide, or lying on a slab in the morgue. And that's no truffle ... er, trifle. But if you carefully compare each specimen you find with the illustrations in a mushroom hunter's field guide, or to the descriptions in this article, you can narrow the possibilities down to a few 'look-alikes.' From there it is a simple matter to make a positive identification by examining the spores under a scanning electron microscope. With the advent of the new portable models, no mushroom hunter should be without one in his field kit.

The Mushrooms

Common Name: Mushroom
Genus, species: Agaricus bisporus
Description: Small white button cap 2-9 cm in diameter, short stalk 1-3 cm long.
Habitat: Can be found in clumps, usually beside broccoli or cauliflower in produce section.
Edibility: Edible but boring.
Look-alikes: They all look just about the same. Some are bigger, some are smaller.

Common Name: Disk mushroom
Genus, species: Agaricus frisbus
Description: Large disc-shaped cap, 24-36 cm in diameter, turning under at the margin; brightly colored, usually red or blue. Stem is usually very short or missing altogether.
Habitat: Open areas such as parks and playgrounds. Sometimes may even be found on rooftops.
Edibility: Edible but very tough. Some have described its texture as almost like plastic. Those who eat this species regularly recommend parboiling it for several days. I myself have had little luck making this one palatable.
Look-alikes: Agaricus hubcapius

Common Name: True Molar, Dog Mushroom
Genus, species: Morchella swellfella

Description: Round, disc-shaped cap sitting on a tubular stalk.
Habitat: Deep in the woods.
Edibility: Edible and choice.
Look-alikes: Gyromitra biteya
Comments: This species is particularly delicious and as such is much sought-after. It has a robust flavor and a meaty texture, making it an excellent choice for meat-craving vegetarians. It is sometimes called the mushroomer's best friend.

Common Name: False Molar
Genus, species: Gyromitra biteya
Description: Round disc-shaped cap sitting on a tubular stalk.
Habitat: In the woods.
Edibility: Deadly!
Look-alikes: Morchella swellfella
Comments: Onset of symptoms occurs from 12 to 48 hours after ingestion. The first symptom is the loss of the ability to pronounce the letter 'F', followed by mild to severe death. Sometimes a positive identification between M. swellfella and G. biteya can be made on the basis of whether it was found in the woods, or deep in the woods.

Common Name: Squirrel Catcher, Venus Squirrel Trap
Genus, species: Amgoneta grownsqurl
Description: Large bowl-shaped cap, 20-30 cm in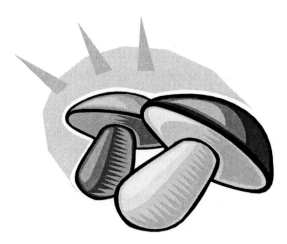
diameter, partially filled with slimy, acrid fluid. Slender stalk protrudes slightly above liquid level. Knob at end of stalk resembles acorn.
Habitat: In the woods.

Edibility: Edible, but who would want to?
Look-alikes: Amgoneta chipmunk
Comments: A squirrel sees the acorn, climbs in to fetch it, slides down the slippery sides of the cap into the liquid and dissolves. Yum! Most squirrels give this species a wide berth.

Common Name: Screamer
Genus, species: Agaricus decibellus
Description: Round disc-shaped cap sitting on a tubular stalk.
Habitat: In the woods.
Edibility: Edible.
Sound-alikes: Fingernails on blackboard, unhappy rhesus monkeys.
Comments: This species has developed one of the best defense mechanisms short of being poisonous. It screams at the top of its gills when you try to pick it. It is mainly because of this species that no experienced mushroom hunter is without ear plugs in his field kit.

Common Name: Reviled Amanita
Genus, species: Amanita yodeadsucca
Description: Round disc-shaped cap sitting on a tubular stalk.
Habitat: Anywhere it wants to be.
Edibility: NO WAY!
Look-alikes: Agaricus bisporus, Agaricus frisbus, Morchella swellfella, Gyromitra biteya, Amgoneta grownsqurl, Agaricus decibellus and others.
Comments: My mother taught me that if you can't say something nice, don't say anything at all, but I am going to make an exception in this case. This is without a doubt the worst, the most despicable, malevolent, conniving, malicious, malignant, abominable, mean, and generally most thoroughly disgusting mushroom known to man. Not only does it contain several toxins for which there are no known antidotes, but it has also been known to trip up and bite hapless hikers. There are several cases on record of hunters contracting cancer from just looking at one too closely. Just before he died, Dr. Rollo Moss commented, "It seems to be sneering at me." Apparently some people can eat this species with no ill effects, but I cannot recommend it.

How to Survive a Robot Uprising: Tips on Defending Yourself Against the Coming Rebellion

excerpted with permission from the book by Daniel H. Wilson, PhD. Bloomsbury, 2005

If popular culture has taught us anything, it is that someday mankind must face and destroy the growing robot menace. In print and on the big screen we have been deluged with scenarios of robot malfunction, misuse, and outright rebellion. Robots have descended on us from outer space, escaped from top-secret laboratories, and even traveled back in time to destroy us. The cultural icon of the killer robot goes back almost as far as the notion of the "mad scientists" who supposedly create them. Even the word robot has ominous roots. It is Czech for "laborer" and was coined in R.U.R. (Rossum's Universal Robots), a play produced in 1920 in which robots revolted and destroyed all humans.

Today, scientists are working hard to bring these artificial creations to life. In Japan, fuzzy little real robots are delivering much appreciated hug therapy to the elderly. Children are frolicking with smiling robot toys. It all seems so innocuous. And yet how could so many Hollywood scripts be wrong? How could millions of dollars of special effects lead us astray? So take no chances. Arm yourself with expert knowledge. For the sake of humanity, listen to serious advice from real robotics experts. How else will you survive the inevitable future in which robots rebel against their human masters?

Humanoid Robots

If imitation is the sincerest form of flattery then humanoid robots (often called androids) serve to help mankind pat itself on the back. The idea of a man-made man is nothing new (think Frankenstein) and science fiction is filled with humanoid heroes and villains, from the merciless T-800 model Terminator to the noble (if chatty) golden C-3PO of Star Wars fame. Outside of movie theaters, real-world scientists are methodically building robots that emulate the human form. Evoking scenes from a cheap horror flick, different laboratories play host to various disembodied prototype android heads, legs, and arms.

As a rule, humanoid robots are bipedal, standing on two legs, and can range anywhere from the size of a cat to the size of a house. Walking like a man, called bipedal locomotion, is no piece of cake – just ask any one-year-old. Early humanoid walkers had bulky gear motors attached to each moving joint, resulting a jerky robotic gait. Contemporary humanoid robots are more dynamic. The Honda ASIMO, which looks like a child-sized astronaut, is remote-controlled and can traverse stairs and hallways at a whopping top speed of nearly 2 mph. The Sony QRIO is more inquisitive; standing roughly as tall as ASIMO's hip, QRIO can move on its own, hold short conversations, and scramble back to its feet after falling down (or being pushed). In contrast to robots, we human beings walk without maintaining constant control of our limbs, in a state more akin to controlled falling. Next-generation humanoid robots copy our passive dynamic gait; instead of placing each step precisely, they smoothly swing their synthetic limbs. The resulting motion is more energy-efficient (it exploits the momentum of the step) and more resistant to failure (there is room for uncertainty in the terrain).

Physically, humanoid robots are starting to amble, trot, and tippy-toe, but how fast are they mentally? One way to side-step this tricky issue is to control a robot through telepresence. With telepresence, a person feels as though they are the robot by controlling the robot's body and seeing through its eyes. Human-shaped robots are easier to manipulate because there is a 1:1 mapping between man and machine. NASA has developed the RoboNaut, a functional robotic replica of a human from the waist up (and an eerie, jointed tail from the waist down). Through RoboNaut, a technician in Florida can replace light-bulbs on the outside of an orbiting space shuttle. For now, the human is the puppet master, but once humanoid robots are gifted with their own intelligence what fiendish goals might they pursue?

Humanoid robots currently stalk only the hallways of research laboratories, but they have a huge potential to act in the real world, as all-purpose machines that can take care of life's little details: cooking, cleaning, and chasing kids off of your lawn. As their physical abilities outpace humankind, however – they will run faster, jump further, and lift heavier loads – the threat to our species will be revealed, especially when a humanoid robot steals your deer-hunting rifle and drives away in your pickup truck, firing wildly into the night.

How to Escape a Humanoid Robot

One minute you are strolling across an empty parking lot with arms full of groceries and the next minute 2 tons of steaming bipedal man-bot is bearing down on you. A humanoid robot may look like you, but it is probably faster, stronger, and much better at chess than you are. Drop the groceries; it's time to learn how to escape.

Run toward the light. Vision sensors are confused by sudden changes in lighting conditions. Forcing the robot to follow you into the Sun may slow down its pursuit.

Find cover. If there is none, find clutter. Put obstacles between yourself and the robot. Cover is anything that can both protect and hide you. If there is no cover, use clutter, anything that hides you and befuddles robot vision.

To save a comrade: first merge, then separate. Run to a comrade, deliver a quick bear hug and then dive in a random direction. A vision-based target tracker might temporarily lose track of your identity during the hug, especially if you are wearing similar clothing. You can gain precious seconds while the tracker re-acquires its target.

Don't run in a predictable line. If you follow a simple velocity trajectory, it will be easier for a robot to track your progress, even through significant clutter. Zig-zag erratically or, when hidden from view, change direction suddenly in order to throw off predictive tracking systems.

Use rough terrain. A humanoid robot can run faster and for far longer than you can. Take pride in your primate heritage – humanoid robots are not as good as you are at scurrying over walls, climbing hillsides, or clambering over and under parked cars.

Find a body of water. Most robots will sink in water or mud and fall through ice.

Find a car and burn rubber. Theoretically, a humanoid robot could sprint as fast (or faster) than an automobile, but the resulting heat and stress would likely overheat or injure the robot pursuer.

Modular Robots

In concept, modular robots are made up of thousands or millions of tiny modules, just like cells in living things. The similarity to animals stops there. Modular robots are infinitely more adaptable, able to reconfigure into almost any shape imaginable. If every module is identical, a damaged modular robot can self-repair, shedding broken modules or easily absorbing new ones. Collected together, the modules may resemble a pile of sand or a pool of mercury, but when the pieces communicate there is no limit to the nightmare forms a modular robot might assume.

While current modular robots hardly resemble the terrifying (and sexy) T-1000 Terminator, research is well underway. The Crystalline Robot from Dartmouth University looks like an animated set of children's blocks, composed of dozens of brightly colored plastic cubes (2 inches squared) that expand and contract in 2 dimensions (on a flat surface). The cubes communicate with one another via infrared light; together they creep across the floor as one entity, parting and flowing around obstacles like slow-motion water. In simulations, a newer version can operate in 3 dimensions to engulf a smaller object and manipulate it from the outside, using hundreds of modules like tiny fingers. In California, researchers at the Palo Alto Research Center developed the Polybot, which uses hinged chains of cube-shaped modules to change its form from a serpentine shape to a gangly spider for stepping over obstacles. A recent competition saw dueling Polybots assume forms ranging from humanoid to serpentine.

Possessed with remarkable malleability, modular robots offer great promise, fraught with formidable challenges. Currently, modules are small in number, large in size, and must laboriously search for each other before docking at an agonizingly slow pace. Complicated locomotion and manipulation activities are limited to computer simulations. Still, advanced prototyping methods are creating ever smaller modules that communicate and change form ever faster. In the distant future, modules could shrink to the size of dust specks or even become microscopic nanorobots. Modular robots are ultra-complex, nearly unstoppable, and have the potential to eventually replace all other robot forms.

How to Stop a Modular Robot

Modular robots are especially insidious – they can assume any form and they can self-repair almost instantly. To defeat this type of robot, whatever shape it takes, you must act fast and think faster. During an attack you may only get one opportunity – so learn how to take it. Save your bullets. Bullets will damage only a tiny fraction of the robot's body. However, some modular robots may be carrying tools or have specialized modules designed for specific tasks. These are worth destroying because they cannot be repaired immediately.

- Don't bother hiding. You can run but you can't hide when your pursuer is capable of pouring itself through a keyhole (just ask Sarah Connor).
- Trigger a transition or repair phase. A modular robot is most vulnerable during reconfiguration or repair. During this time each module must communicate and maneuver into a new position. Modular robots change form to solve new problems. So trigger a transition phase by presenting the robot with a new problem – like parking your Geo Metro on top of it.
- Divide and conquer. If the modules are unable to communicate, there can be no consensus. During a transition phase, parts of the modular robot will lose shape and flow into new configurations. Focus your attack on these vulnerable areas, which may look like flowing liquid or sand.

- *Scatter the modules.* Kick, slap, shoot, or do whatever you have to in order to separate the modules and keep them separated.
- *Coat the modules.* Use any available liquid. Thick, dark syrupy liquids will work the best. Opaque molecules will stick to each module and block communication pathways.
- *Mix up the modules.* If no liquid is available, throw any kind of foreign matter into the reconfiguring robot. Bits of dust, leaves, or shrapnel will have to be expelled before the robot can assume its final shape.

Your Body is a Numberland

Lawrence M. Lesser, PhD, University of Texas at El Paso

☛ goes well to the tune of John Mayer's *Your Body is a Wonderland*

Your figure's got beauty
With bilateral symmetry,
And your span from head to toe —
Your navel makes golden ratio!

Your skin's almost 2 square meters:
Finding area was never sweeter!
Your fingers curl a right-hand rule:
Math extends beyond our school

CHORUS:
(And if) you want math — just do it!
Explore a relation — we'll prove it!
Take all our problems —
and solve them,
Even if it takes a while:
Your body is a numberland —
Put down the calculator,
we'll do it by hand.
Your body is a
numberland ...

The number of your
fingers
is our number system,
And counting
builds harmony and
rhythm
That your spiral ear
turns into sound
While the lens of your eye
turns me upside down!
(REPEAT CHORUS)

Angel Valdez

Nature versus Nurture in Color Discrimination

J. G. Mexal

How often have you asked "Is that color pink, or peach, or salmon?" If you have asked that question even once, you are probably female. If you don't know the difference, you are probably male.

The difference in color discrimination between the genders appears to lie in the Corpus Crayola® region of the brain. In the female brain, this region is highly convoluted, indicating a high degree of development. However, the male Corpus Crayola is underdeveloped almost to the point of lacking any noticeable features.

This might explain why men can identify only about 8 colors. Adding the modifiers "light" and "dark" brings their recognition up to 24 colors; e.g. "brown", "light brown", and "dark brown". However, the female Corpus Crayola can identify up to 24 variations of brown alone, e.g. wheat, tan, sand, khaki, brown, cocoa, sable, chestnut, mocha, burnt sienna, mahogany, sepia, walnut, seal brown, and chocolate, to name a few. (The author, a male, obviously had female help in developing this list, but could not define these shades much beyond "BROWN"!)

The Corpus Crayola lies at the base of the brain, near the brain stem, indicating early development in the evolution of human behavior. However, it is also connected to the optic nerves, and the frontal lobe where emotion resides. This would indicate its involvement in visually stimulating activities, such as shopping, and visiting art museums – obvious gender-specific activities.

The differences in the development of the Corpus Crayola do not separate exclusively along gender lines. Some men have well-developed color distinction skills. Generally, interior decorators, fashion designers, hair stylists, and baseball switch hitters have well-developed Corpus Crayola regions. This leads to the hypothesis that the differences in development of the Corpus Crayola are not under genetic control (Nature), but rather under environmental control (Nurture).

My suspicion, based on personal experience, is that most male children are given only the small box of crayons, with just 8 colors. On the other hand, female children routinely get boxes of 64 or even 128 crayons. Thus, the visual stimulation that might lead to development of the Corpus Crayola is set at a very early age.

There may be a reproductive strategy to limiting the color selection of males. Young boys often hear their mothers say "Honey, those don't go together. Let me help you pick another shirt." This becomes so engrained in the male brain that, later in life, they may instinctively select mates who can continue to dress them in the morning. Thus, selecting a breeding partner similar to their mother is a function of a box of crayons.

☛ **Can you distinguish beige from taupe?**

Sand vs Stars — The BS9 Project
Dale Lowdermilk, Santa Barbara, California

How Many Star Names Can You Count?

According to the *Sky Atlas 2000.0*, just after midnight on a perfectly calm, cloudless, windless, moonless night, far away from city light pollution, you could see the 8,769 stars of magnitude 6-point-something or brighter. This represents the WORLDWIDE total of stars visible to the unaided eye, if you're lucky enough to have perfect vision. By living in the Southern Hemisphere, somewhere in the Australian outback, your total would be just over 4,600 ... if viewed from the Mojave Desert during winter months, the total would be about 3,500.

The SuperNova Business of Selling Star Names

Not to pop your astronomical "bubble", but most stars don't have names ... they have cryptic letters or numbers representing their category, character, or coordinates. To tell someone that you are going to name a star after them is equivalent to selling one bottle of celestial snake-oil over and over again to different individuals. More bluntly, all the best stars (Rigel, Sirius, Altair, Betelgeuse, Polaris, ...), and the constellations, have already been named. An internet Google® search conducted in March 2003 for "name a star for someone" resulted in less than 100 "hits". That same search, conducted in May 2005, resulted in 44,300 "hits", which may mean there are now more companies selling star names than there are visible stars. This dramatic growth in astro-vanity may be a direct result of declining Math and Science scores seen among high school seniors in recent years. How else could we explain the behavior of people willing to trade their money for a piece of paper containing valueless information?

From Big Bang To Celestial Fraud

According to the International Astronomical Union (IAU), the only official organization authorized to assign a name or number to celestial objects ... "As an international scientific organization, the IAU dissociates itself entirely from the commercial practice of 'selling' fictitious star names or 'real estate' on other planets or moons in the Solar System. Accordingly, the IAU maintains no list of the (several competing) enterprises in this business in individual countries of the world." The practice of huckstering "star names" is a pure commercial hype with no basis in fact or Science.

High-resolution 2400 dpi scan of only the finest hand-selected sand granules from Santa Barbara, California, and printed with noncarcinogenic inks on recycled biodegradable paper. No animals were harmed during this printing. Price $200.

"Stars-in-a-Jar" — Better Than Obscure Coordinates

Instead of being ripped off by supernova scams, consider offering your loved ones something tangible and permanent, such as a grain of sand. We can sell you an actual piece of valuable beach-front Santa Barbara real estate, a certificate suitable for framing, and genuine GPS coordinates to help you find your place in the Universe. As an old tradition handed down from ancient civilizations, "sand-naming" is less offensive, less fraudulent, and more integrated into Nature's Plan (and

Plan B), because we say it is. One legendary tale from an aboriginal wise man reminds us to think of it NOT as sand but as STARS-IN-A-JAR. (Just don't try to drink them during hallucinations.) The gem-quality sand grains you purchase from The BS9 Project have been harvested, inspected, and packaged by professional sand-reapers who have been fully certified and licensed by CRAP (California Regional Attenuation Processors.)

What Is The "BS9 Project" And How Can I Spend My Money?

Our parent company, ISGR (Intergalactic Sand Grain Registry), and TSMF (Terrestrial Silicates Microplanetesimal Foundation) have joined together to form "Buy Santa Barbara Sand Before Someone Blames U4 Getting Surreptitiously Bluffed w/Stellar Balderdash & Simply Brilliant Strategy"... aka the BS9 Project. Our geologists, astronomers, tourism, and financial experts are conducting a long-term (and very expensive) study to investigate the phenomena of aberrant marketing techniques (AMTs) and inexplicable spontaneous purchasing behaviors (ISPBs).

Waste Your Money On a Point Of Light, or Invest In Pure Santa Barbara BS?

Don't squander $30-$75 of your hard-earned money thinking you'll honor someone with a white dwarf, red giant, globular cluster, or high-speed transient object which is invisible except through the Hubble Space Telescope. With only 8,769 visible naked-eye stars, there's a good chance that if you can see your "unique stellar object", it has been sold and resold hundreds of times by cosmo-shysters. The BS9 Project is an ecologically-sound, non-corporate, ethically-reconciled, quasi-commercial venture which does not pollute the wetlands, disrupt salamander breeding grounds, or traumatize sensitive insect larvæ. Purchasing sand will make you feel good from the checkbook out. The BS9 Project can be all things to all people. Our motto is, "Mirth In The Balance ... No Pain, No Grain ... Leave Your Money In Our Sand Castle Then Go Home".

Our competitors, in the fraudulent "star-naming" business, complain that selling grit is just as unethical as selling invisible stellar coordinates ... but they are

WRONG. The BS9 Project is absolutely TERRESTRIAL. There's a huge difference between selling precious particles of REAL silicates, and selling gullible clients ET's street address. Santa Barbara is the only community in America with genuine $1200 litter disposal receptacles (garbage cans) and $2000 cornstalk-shaped street lights which illuminate a minor thoroughfare frequented mostly by the drunk and homeless.

Sand Grain Memorial Selection--$75

IN MEMORIAM

This grain of sand #8059696217 has been registered in the name of

J.A. NeDoe

World's Greatest Riter
Most Anonymous Politician
John's Best Friend

Our sand doesn't stink ... the sedimentary elite granules from our invaluable beach environment are True Hands-On Naturalgasmically Gendered Microplanetesimals Registered In Transmogrification (THONGRIT) Each and every granule is as unique as a Michael Jackson home video, and guaranteed non-synthetic. (Beware of imitations. Accept only BS9 Project certified beach sand.) No chemicals, nuclear fusion, or telescopes are necessary to memorialize your loved one when you trade your money for PURE SAND instead of a star name.

Your investment in Santa Barbara beach particles will help the BS9 Project analyze why people are willing to squander their cash under the guise of love. Unlike gold, an investment in pure sand makes a powerful political statement against unilateral violent nuclear fusion reactions occurring elsewhere in the cosmos. Our sand requires no batteries, comes ready to use, and you won't need to purchase a pair of binoculars or an expensive telescope manufactured in some remote sweat shop by a 10-year-old earning $2 a day. Your consciousness will be elevated to harmonically converged levels, your money will be in our pockets, and your heart will be at peace. Sand can be inherited TAX-FREE by your heirs through proper estate planning. Our BS9 Project attorneys will gladly help you with your inheritance issues or questions regarding interstate transfer of sand.

Every speck of silicate we sell has been endorsed by the National Committee to Preserve Famous Original Beach Sand and the Santa Barbara International Sand Festival. Finally, we offer a 10-year guarantee against shrinkage for each grain. Extended warranties against cracking, crumbling, or bacterial toxicity can also be purchased.

Which Came First — the Sand or the Stars?

The BS9 Project guarantees that your grain of sand is personally inspected for color, symmetry, radioactive contaminants, and spectrographic pedigree. We further guarantee that the name assigned to your individual grain of sand will never be reissued to someone else. Don't be confused by imitations such as gutter grit or inferior particles of soil from the Mojave or Sahara deserts, which could have been trampled upon by sweating tribesmen or exposed to camel dung. Nor will your sand have been imported from polluted or disease-ridden beaches of Bangladesh or San Diego. Only BS9 Project sand is pure, clean, and unique. NO OTHER SAND GRAIN REGISTRY CAN MAKE THIS REMARKABLE CLAIM!

Using inductively-coupled plasma-atomic-emission-spectrometric-analysis (ICPAESA) and the "taste test" we can show conclusively that our pure, natural beach sand comes directly from the finest beach in the world.

For an additional $35 we'll send you a hand-drawn sketch and certificate of authenticity describing the exact "zone" where your personalized grain of sand was harvested. We promise to consider publishing your name along with a series of random numbers in a book which will be registered in the US Copyright Office and might make the *New York Times* "Best-Seller" list.

Personal Achievement Award--$25

This grain of sand #4084640073 has been registered in the name of:

Ms. Bertha Butts

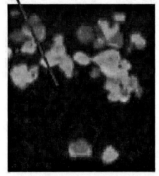

Congratulations on your Graduation.
Release From Prison
Dramatic Weight Loss

From The Feet of The Elite — Pure Grit Lasts Forever

After years of arduous searching for PERFECT granule candidates, one extraordinary and very expensive Southern California location meets or exceeds all requirements. In Santa Barbara, California, there exists a superb coastal zone where famous movie stars secretly come to spawn and get their toes gritty. During these long walks (and sometimes drunken-stupor-crawls) your own personal grain of sand is exposed to sweaty body parts and feet of the rich and famous. These superior sand particles often touch the finest looking bikini babes on Earth as they stretch out to toast their epidermal layers. You might even end up with a grain of sand that Oprah stepped upon! This is truly the beach sand of the stars ... REAL movie stars mixed with REAL sand, not just abstract cosmic coordinates scribbled of piece of paper.

According to recent sky surveys, there are more stars (visible with telescopes) in the universe than there are grains of sand on all the beaches and deserts of the world. Using a secret mathematical formula (included with your purchase of $200 or more) you can prove to your friends that your "grain" is more unique than some stupid amalgamation of hydrogen and gravity boiling in the bleak darkness of eternity. Our guiding philosophy here at the BS9 Project is to NEVER INVEST IN A BLACK HOLE.

It's a complete waste of time to support untouchable fusion reactions occurring somewhere in the cosmos. Don't dishonor the memory of your favorite spouse, friend, lover, coworker, or pet with a frivolous star name that will just blow up in your face in a few million years. Instead, give them something eternal and tangible ... genuine Santa Barbara Beach Sand. Everyone here at the BS9 Project is proud to be a part of this unique business/Science venture. When choosing the perfect grain for your favorite someone, remember that, like sand on steroids (diamonds), grit can also last forever.

Each specimen from the BS9 Project is individually and meticulously hand selected (using tweezers to avoid contamination) and mounted on a glass slide for easy viewing from any light source. Some people prefer to study their sand grains through an electron microscope, noting subtle changes in color or shape due to atmospheric conditions as a result of the Administration and global warming.

Don't rely on some abstract, impossible-to-locate stellar coordinates, when you actually visit the home of your own little silicate crystalline mass in person. When visiting Santa Barbara, be sure to take the BS9 Project Grid Location and Historic Sand Identification Tour.

Sand Investments — The Wave of the Future

Here's the way it works. For as little as $25 you get a genuine grain of sand named after your favorite lover, friend, spouse, child, coworker, pet, rock-star, politician, hallucination, top-achieving employee, or college graduate.

Unnamed sources at high levels of city government have indicated that secret plans are being negotiated to erect environmentally safe and biodiversely constructed fences and location markers. These will be constructed to protect the "purity" of Santa Barbara's ingrained (possibly inbred) silicate heritage. Harvesting these extraordinary little pebbles occurs only during the dark of the moon and highest winter tides, to coincide with the grunion runs. Some officials predict that "sand-harvesting" could become the next major industry along Southern California beaches. This would produce astounding tax revenues, which could be used to promote further tourism and sand consumption. However, others fear that cheap "black-market" sand from Mexico could begin flooding the market. Here at the BS9 Project we know these petty little facts are as important to you as your money is to us. For a limited time, you may still visit these rare, unspoiled sand-nurseries. Today, these zones are free, but with California's perpetual budget crises, this could soon change. Some BS9 Project Committee members have proposed charging a nominal fee based upon the number of grains of precious sand that stick to the feet of tourists as they enjoy their beach walks. This, in conjunction with strict sand-smuggling fines and full-body searches, will deter those trying to make a few illegal bucks on Santa Barbara's seaside commodity.

Guided tours conducted by unemployed jitney drivers will show you exactly where to find future "plot layouts" and fast-track expanded sand-grain harvest areas (wholesale buyers only). Special grids will be set aside for low-income purchasers. (See photo)

At The BS9 Project, What You See Is What You Grit

If you order a family-sized bucket of Santa Barbara Beach Sand grain today we'll include:
1. A list of famous people who probably walked upon it.

2. *101 Facts About Sand* (Why Santa Barbara sand really is different from crushed rock, dust, dirt, grit and grime.)

3. *10 Things You Can Do With Sand* (Some cultures wear it, trade it, meditate upon it, and even use it for barter.)

4. *50 Waves To Start a Conversation With Beach Sand ... The Beginner's Book of BS*

5. Instructions on how to make your own hourglass.

6. New non-mathematical methods to calculate the million grains of sand in a cubic inch (Impress your friends!)

7. How ancestral sands from Santa Barbara began the modern environmental movement. News clippings from the Great Cash Spill of January 1969. Explains how "Big Oil" sparked outrage by killing millions of sea mammals, birds, lobsters, and fish. Includes best selling beach-tar-removal booklet *Your Grain Was There*

8. *12-Step Program To Sand Withdrawal* (Just in case things get out of control)

9. Accurate map indicating where three (3) Elvis sightings occurred on Santa Barbara's East Beach in 1992.

10. The BS Sports Guide (Volleyball and frisbee. Includes excerpts from *Professional Wedgie Techniques Using Sand*)

11. *Introduction to Sand Castle Designs* (This handbook includes many famous "Sand Related Insults")

12. *How To Get Out of a Relationship Using Pure BS* (Invaluable tips for young single males.)

13. *Poems About Sand* (Beautiful readings from sandaholics around the world)

14. *Sand and Sex Throughout History* (Excellent "how-to" book with Kama-Sutra-like diagrams)

15. *50 Greatest Blond Sand Jokes* (A collection of wild and wacky items from the best internet sand sites.)

16. *Cooking In Sand* (Beach meals you can prepare on a stick)

17. *Tips for Safely Removing Unwanted Beach Sand From Most Body Orifices.* (Written by medical experts)

18. *Dale's Field Guide to Sand Watching* (Color illustrations of more than 250 different grain, shape & size categories)

19. *Nostradamus Knew His Sand* (12 pages of amazing predictions that prove sand is real)

20. *Fallwell's Biblical Prophecies About Sand* (Genesis 22:15-17, Genesis 41:49-52, Matthew 7:26, Psalm 139:17-18)

21 *Grit In Their Eyes* (Proven methods of self-defense. Learn how to subdue aggressive foreigners by using subsonic sand particles and spit.)

22. *Turning Sand into Glass* (Things to do with your sand when it misbehaves)

23. *The Mother of All Rat Holes — Personal Memoirs of Saddam Hussein.* Follow the last days of this notorious sandman. Transcribed from his own blood-and-urine-stained notes. Learn how sand became his only consolation, then his enemy as it collapsed around him. Read in his own word(s) how grains from his own personal hell betrayed his location. Learn how he used the internet to locate the ideal "hole-in-the-ground" facility. Discover ways to improve your own little sand castle and how to differentiate "friendly" particles from "enemy" grit. Chapter 7 of this historic document details how the infamous dictator befriended 2 rats and a cockroach, only to eat them a week later.

Discover how Saddam tried to hide his favorite grains of sand in his hair by camouflaging his scalp with chocolate, and how US troops discovered his own personal grain of Santa Barbara Beach Sand hiding cleverly between 2 decayed lower molars. This became the famous oral examination by Army doctors as they discovered the valuable grain on video tape before a world-wide audience. When captured, Saddam attempted to "negotiate" his release by using this precious nugget purchased from the BS9 Project in 1999.

It's Not Just A Souvenir — It's A Mineral To Remember

Although it tastes funny, sand has no calories, is not radioactive, will not become a supernova, and no one ever went blind staring at a grain of sand. How many stars can make those claims?

Stop supporting long distance nuclear fusion and supernova overreactions by buying a safe and sane piece of beach grit for someone special. Dispersal of these intergalactic presolar microplanetesimal silica granules (IPMSG) can make the Universe a better place.

For $34 we will name and send you a photograph of your personal sand grain. Overnight delivery is possible, making Santa Barbara Beach Sand the perfect last-minute gift idea. We'll provide a detailed satellite map, like the Beverly Hills celebrity home maps, so you and future generations of beach-goers can locate your own personal grain of sand. What better excuse for that beach vacation than "visiting" the site. Each grain of sand is as unique as a snowflake ... no 2 are exactly the same ... sand pebbles are as special as the individual or the pet you wish to memorialize.

Our Intergalactic Sand Grain Registry Gift Shop offers gift certificates and a variety of 8" x 11" official BS9 Project frames.

Send us your credit card order today and we will give you THREE (3) grains of sand for the price of one. Order more than 10 grains of sand and we will include a magnifying glass to view them in detail. For an additional $50 (per grain) we will include a complete chemical analysis with a detailed and beautiful spectrographic chart suitable for framing. If you do not have room to store your memorial sand grain, we will for a nominal fee ($200 per year) place it in a cryogenically secure clean room until you decide what to do with it. Perhaps you will want to return it to its place of origin on Santa Barbara's exclusive East Beach ... we can help. Our highly motivated BS9 Project accountants can show you how to use estate planning to assure that your eternal sand grain is passed to your heirs without probate or taxes, and how wholesale and quantity purchases can help your favorite charities or foundations.

Family Portrait: $200
Memorial Selection: $75
Personal Achievement: $25
Unregistered Bucket of Sand: $10
Unregistered Teaspoon of Sand: $5
Special Orders: $50 per grain

Name-A-Grain (Laser Imprinting) — GRAFFITI THAT LASTS FOREVER!

Don't lose your sand during a home burglary or to a pickpocket, get it imprinted with your name (visible only under a microscope). If you tend to forget where you have placed your precious sand, this will help you find it. (If you have found a personalized grain of sand, please return it to your nearest police station or law enforcement authority, or report it by using our on-line registry at www.notsafe.org (attention BS9 Project).

Notes & References

Footsteps in the Sands of Time (H. W. Wrongfellow)
New International Version Bible (Genesis, Exodus, Abraham, Moses, Buddha)
Sand and Sexually Transmitted Diseases (Published by Trojan Corp., 1996, chapter 7, et al.)
Sunny Beaches of the Rich and Famous by Santa Barbara Visitors Bureau
Who's Who In The Ground © 2001 by Saddam Hussein, and *Dictator in His Underwear* 5/25/05 *LA Times*
Who Let The Sand Out ... Ohh Ohh. ™ 2004 Dalina Michaels@Yardi
Don't Mix Grit and Upgrades by Michael Lowdermilk@fastmac .com, Lyrics "Clean Up This Mess" & "I'm Selling Your Sand Collection" by Pierina Guadagnini, p 471, Chapter 34
Einstein — What Did He Know, And When Did He Know It? By Dale Lowdermilk 805-969-6217

Additional Funding Provided By Name-A-Grain Foundation, National Endowment for the Farts, Federal Registry of Gullible Gift Givers, the Editorial Bored of **The Journal of Irreproducible Results**, and readers like you.

☛ **Selling star names is a scam: the sellers don't own what they sell. What other scams are ripe for satire?**

Acoustic Oscillations in Jell-O®, With and Without Fruit, Subjected to Varying Levels of Stress

© James Stanfield, co-founder, The Institute for Further Research

The test setup: A standard preparation of cherry Jell-O was mixed according to package instructions, then divided into 2 Pyrex® bowls, A & B, setting aside a certain quantity for make-up. A ripe, but not overly ripe, banana was sliced into 1 cm ±1 mm segments and added to bowl B. Jell-O from the reserved make-up liquid was added to bowl A to bring its level up, to equalize the volume in the 2 bowls. Both containers were then chilled slowly to induce a phase change. The

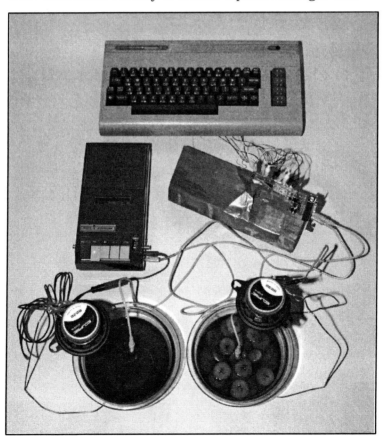

reason for the slowest possible chilling rate will be revealed in due time.

Samples of each of the following sounds had been recorded at a standardized 50 dB: white noise, soothing classical music, elevator music, hard rock, bagpipe music, rap, and, finally, insulting phrases spoken in a scolding tone. Acoustical probes were inserted in the center of each bowl. The bowls were then arranged directly beneath a set of matched electrostatic speakers such that they would receive identical acoustic stimulation.

The probe leads were then plugged into the dual channel inputs of a data-acquisition card, which had been jury-rigged to a Commodore 64 computer for graphical analysis and data reduction. I must say, at this point, that I am disappointed at the shortsighted-ness of my colleague and co-founder of the Institute. He insisted that it was his weekend to use the Commodore 128. Sometimes, I suspect that he is deliberately trying to thwart my scientific research. But I digress.

The author hypothesized that the aforementioned slow cooling rate would induce a phase change to a metastable quasi-crystalline state between the liquid and solid phases, which would semi-periodically tessellate the volume of the bowl. These quasi-crystalline units would then behave as cellular automata. Since each bowl contains several moles of solution there would be several Avogadro's numbers of these quasi-crystals, and the quantum computing power contained in this self-organizing complexity might very well be immense!

Awareness (thought, or at least, thought-precursors) might spontaneously arise as a metaphenomenon of this self-organized complexity, which could then perceive subjective differences in the stress levels of the various acoustic samples and react accordingly. Primitive reflexes might evolve.

It is presumed that there is a critical computational mass that is required for this to occur. It was decided that the fruit would be added to this experiment on the off-chance that it would act as an organic catalyst and thus reduce the amount of Jell-O needed to achieve this critical mass.

After many runs utilizing each acoustic sample in turn, the graphs were compared for differences. (My money was on the rap music.) No differences were found between the various acoustic samples within a given bowl. However, between bowl A and bowl B there was a flattening (lower peaks, higher troughs) in the standing wave pattern evident in the bowl of Jell-O with the bananas. This artifact could be due to a damping effect because banana is much less 'jiggly' than Jell-O.

So far the results are inconclusive. At this point, whether critical mass can be achieved with 100 gallons, 1,000 gallons, or even 1,000,000 gallons of Jell-O is anybody's guess.

Some of the possible reasons that the experiment did not go quite as planned are as follows: 1) The experimental setup did not have the required sensitivity. 2) The Jell-O was in an uncooperative or indifferent mood (like a cat). 3) Jell-O is an inanimate substance. 4) Awareness does not spontaneously arise out of self-organized complexity. 5) Consciousness may very well require a coherent quantum state within the medium in which the Platonigenic Field is acting. In this case, the author must postpone further experimental trials until Bose-Einstein-Condensate-flavored Jell-O becomes available.

And, since the grant proposal requesting funding from the NSF was rejected, I feel that I am not under any obligation whatsoever to come to any scientific conclusion.

A BILLIONTH of a second is such a short time that light only has time to go a FOOT!

I bought one of those new Radio Controlled Atomic Clocks. They're supposed to be accurate to 1 nanosecond. I opened up the box and the darned thing was ALREADY a nanosecond slow. I called Tech Support. They told me to stand closer. It WORKED!

— Norm Goldblatt, www.NormGoldblatt.com

More Than Plural, Less Than Singular

Norman Sperling, with contributions by Jerrold H. Zar, Pino Della Gatta, Lorna Pollock, & B. Kliban

When we're taught that nouns come in singular and plural forms, we get the impression that those are their only quantities. That's reasonable ... but wrong. Sometimes, plural nouns are pluralized further. And sometimes, singular nouns are taken down another level.

Even More than Plural

"Data" is one of the most commonly misused scientific terms. It's plural. The Latin singular is "datum". Yet many people talk about "this data" instead of "these data". Some even pluralize it: "datæ" or "datas".

"Agenda" are items listed for action. Each item is one "agendum", but since that term is very rare, people have an excuse for not knowing it. The list itself is often regarded as singular, which is then pluralized as "agendæ".

The same phenomenon is found with "phenomena". It is quite phenomenal how many times people regard that plural word to be singular, and quite singular how they pluralize it to "phenomenas".

By that system, practically any word that is plural but doesn't end in "s" can be raised to a higher level by adding an "s" or the appropriate Greek or Latin treatment.

Plural terms that are raised to an even higher level might be termed "suplural".

Even Less than Singular

"Kudos" is the Greek condition of receiving praise or glory. When praiseworthy people receive the praise of which they are worthy, they receive "kudos". Many times, the term is regarded as plural since it ends in an "s". Its singular form must be "kudo", which might be a kudu that does judo — which would be praiseworthy indeed.

People sometimes discuss the number of "lens" in a telescope, as if each curve of glass is a "len".

By the same system, practically any word ending in "s" can be taken down a level by slicing off that "s". "Campus" would become "campu", "crocus" would become "crocu", and "ibis" would become "ibi" — though, ending in "i", that looks like a plural form of "ibus"!

Singular terms that are further reduced might be termed "zero-order".

Zero-order	Singular	Plural	Suplural
	agendum	agenda	agendæ
	aircraft	aircraft	aircrafts
alumnu	alumnus	alumni	alumnis
	bacterium	bacteria	bacterias
bonu	bonus	bonuses	
bu	bus	buses	
cactu	cactus	cacti	cactis
campu	campus	campuses	
	confetto	confetti	confettis
congerie	congeries	congeries	congerieses
corp	corps	corps	corpses
	criterion	criteria	criterias
crocu	crocus	crocuses	
	curriculum vitæ	curriculum vitæs	curricula vitæ
	datum	data	datas
discu	discus		discussion?
eclip	eclipse	eclipses	
	elk	elk	elks
Elvi	Elvis	Elvii	
faux pa	faux pas	faux pas	faux pases
fent	fence	fences	
Filofak	Filofax®	Filofæces	
focu	focus	foci	focis
fok	fox	foxes	
ga	gas	gases	
Goretek	Goretex®	Gortices	
	graffito	graffiti	graffitis
hippopotamu	hippopatamus	hippopotami	hippopotamis

119

ibi	ibis	ibises	
indek	index	indices	
influk	influx	influxes	
Kleenek	Kleenex®	Kleenices	
kudo	kudos	kudoses	
len	lens	lenses	
Lexu	Lexus®	Lexuses	Lexi
mas*	mass	masses	
	medium	media	medias
minu	minus	minuses	minutiæ?
	nebula	nebulæ	nebulæs
ok	ox	oxen	oxens
	phenomenon	phenomena	phenomenas
plu	plus	pluses	
preci	precis	precis	precises
	protozoön	protozoa	protozoas
rendezvou	rendezvous	rendezvous	rendezvouses
serie	series	series	serieses
		sierra	sierras
sinu	sinus	sinuses	Sinai?
	spaghetto	spaghetti	spaghettis
specie	species	species	specieses
speciman	specimen	specimens	
	sperm	sperm	sperms
	stratum	strata	stratas
syllabu	syllabus	syllabi	syllabis
Thermo	Thermos®	Thermi?	Thermis?
	vermin	vermin	vermins
walru	walrus	walri?	walris?
ek	*x*	*x*s or *x*'s	excess?

---

*minus-first-order: ma

☞ **Submit more at www.jir.com**

Keratinaceous Material from *Felis cattus*: A Significant Carbon Sink

Peter K. Ades, University of Melbourne, & Trevor J.Rook, Royal Melbourne Institute of Technology

Abstract

Few scientists doubt the serious effects of rising carbon dioxide (CO_2) levels in the environment. It is a major contributor to the so-called greenhouse effect. We describe a mitigating effect of a previously unconsidered factor: the sequestration of carbon in the hair of domestic cats, *Felis cattus*.

Introduction: Greenhouse gases

The greenhouse effect is caused by infrared radiation from the Sun being trapped by the Earth's atmospheric gases, rather than being re-radiated back to space. Atmospheric gases that trap such radiation are known as greenhouse gases. The main greenhouse gases are CO_2, methane (CH_4), nitrous oxide (N_2O), and the synthetic halogenated hydrocarbons. Of these, CO_2 is considered the major greenhouse gas, responsible for over 50% of the greenhouse effect. The rise in atmospheric CO_2 levels over the 20th Century is well documented.[1] The key contributor to the increase in CO_2 levels is the combustion of fossil fuels:

$$C_mH_n + (m + n/4)O_2 \rightarrow mCO_2 + n/2H_2O$$

The effects of a continuing rise in CO_2 are predicted to result in increased storm and cyclone activity across tropical regions, and changing weather patterns across the entire planet. The major effect, however, may be the melting of the polar ice-caps. This would result in rising sea levels, with coastal areas worldwide being inundated, and low-lying islands being completely submerged.[2] This threatens both the livelihoods and social structures of the peoples of entire nations.

However, the increase in CO_2 levels in the atmosphere can be reduced by the development of "carbon sinks" – forms in which carbon is trapped, hence limiting the amount released into the atmosphere as CO_2. We describe a previously unconsidered carbon sink: formation of cat hair.

Results and discussion

Anybody who has owned a cat, or who has known – however distantly – a cat owner, can testify to the persistence of cat hair. Cat hair is almost entirely keratin, a protein. Since protein is approximately 54% carbon, it is not difficult to calculate the amount of carbon that is trapped annually in feline hair.

First, estimate the number of domestic cats. The human population is about 6,000,000,000. It is safe to assume that there are approximately 1,000,000,000 domestic cats. Each sheds about 5 grams per day, or 1.825 kg/year. That makes 1,825,000,000 kg of cat hair. At 54% carbon, that's 985,500,000 kg of carbon. If that was oxidized, it would have made 3,286,000,000 kg of CO_2. Keeping that in the form of cat hair significantly reduces the world's total generation of CO_2.

Although greater material wealth brings greater consumption of material goods, and hence greater production of greenhouse gases – particularly CO_2 – such wealth enables greater levels of cat ownership, too. As we have shown here, cat hair provides a significant reservoir of carbon that would otherwise be released to the atmosphere. This is an effect not unlike the "invisible hand" of the market as described by Adam Smith[3] – an unexpected benefit of economic progress. It would, however, be inappropriate for governments to mandate compulsory cat ownership. This would be an erosion of individual rights, no matter how important cat ownership is to the future well-being of the planet. Hence we conclude that the environmentally responsible action for governments is to adopt a free and untrammeled *laissez faire* economic system. The modern (pre-Keynesian) "economic rationalist" system described by Milton Friedman [4,5] would be appropriate, as this provides the individual wealth that in turn allows greater levels of cat ownership.

References
1. Commonwealth Scientific and Research Organisation (2001). *Climate Change Projections for Australia.* CSIRO, Aspendale, Victoria, Australia.
2. Newman, J. (1999) *Scuba Diving and Snorkeling for Dummies.* IDG Books, Foster City, California.
3. Smith, Adam (1776). *The Wealth of Nations.*
4. Friedman, Milton (1962) *Capitalism and Freedom.* University of Chicago Press.
5. Edwards, L. (2002) *How to Argue with an Economist.* Cambridge University Press, Melbourne, Australia.

Hokey-Pokey

Darlene Sredl, PhD, RN

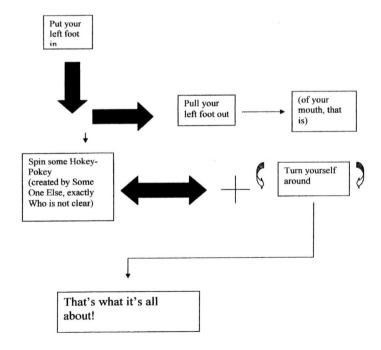

🖝 **Submit more at www.jir.com**

The Real Sleeping Beauty: The Suppressed Story

Rachel Esserman, Endwell, NY

Newly released government documents finally tell the true story of Sleeping Beauty. The material, due to its special nature, remained classified hundreds of years longer than usual. Historians had no idea these papers still existed. They believed the King had destroyed them during his reign. Now, for the fist time, we can learn the real identity of the evil fairy, and the true story behind the spell.

History books have always told the story thus: When the Queen gave birth to a daughter, the King threw a party to celebrate the birth, including as guests many fairies who lived near the kingdom. One fairy was not invited. She came anyway and put a curse on the child. Another fairy changed the curse from a death spell to a sleeping spell. The child was later saved by a kiss from a handsome prince.

The Real Story

The identity of the evil fairy has always been hazy. All attempts to learn more about her failed. New evidence proves she was jilted by Sleeping Beauty's father. Even though he gave her an expensive engagement ring, he never appeared on the day set for their wedding. Instead, the King married a wealthy princess. He thought her a better match since the treasury was low and she came with a large dowry; plus, the Queen's father was a powerful political ally.

Most books claim the invitation to the evil fairy was lost. Others claim the kingdom could not afford any more guests. Neither reason is true. Under the King's order, his knights made sure the invitation never reached its destination. The world did not know of this prior engagement and the King wanted the knowledge to remain secret.

The evil fairy came to the party because she could not stand the disgrace of not being invited. The fairy world had known of her engagement and did not approve. The other fairies felt obligated to accept the King's invitation because he nearly married one of their kind.

The King had an ulterior motive for inviting the fairies. According to the fairy code, any fairy attending a birth celebration must bless the child. This brings increased prestige to both parents and child.

The King had already decided to use his daughter as a pawn in his power struggles. With so blessed a child, he would auction her to the richest and most powerful man in the kingdom, no matter how cruel or evil. Sleeping Beauty's life could have been a horror.

The evil fairy learned of the King's plan early in the day and decided no other should suffer in his hands as she had. Death would be a preferable choice. The rest of the story is the same as in the history books except for one important point. The only man who could break the fairy's spell was one who would be a good husband.

The Cover-Up

The official story published by the kingdom is a cover-up organized by the King and his advisors. Their purpose was to make the King seem like an innocent party by hiding his relationship to the fairy.

The other fairies were horrified when they learned the true reason for the King's invitation. They broke off all contact with humans. The evil fairy became a hermit and nothing more is known about her.

Conclusion

Other books have hinted at the real story but most believed the fairies were part of the conspiracy. The King completely covered his tracks. The few who suspected his involvement had no evidence to support them.

The King, to further his own purpose, destroyed the life of a fairy. He tried to ruin the life of his own daughter. It is now time for history to tell the true story.

Partial List of Documents Found

- Copies of the King's letters (including letters to the fairy and the Queen's father)
- Letters from the Evil Fairy
- Memos from the King's advisors

- Official government edicts
- Minutes of meetings of King with advisors
- King's private journal

Sample Memo
Your Highness,

We'll have no problem discrediting the fairy. No one knows of your plans for your daughter so they'll believe she tried to harm a helpless child. I suggest we leak the story to the press and they will do the rest. Let me know as soon as possible. Time is important.

Other books on the Subject
A Grand Cover-Up?: The Story of Sleeping Beauty by Beverly
   Hartford
Who Was the Evil Fairy?: A Look at Sleeping Beauty by John Bear
Why Wasn't She Invited?: The Mystery of the Evil Fairy by Richard
   Barns
The King and the Fairies: A Hidden Connection? by Jane Hammond
A Fairy Conspiracy: How Much Did the Fairies Really Know? by
   Peter Zander

☛ ***Wicked*** **takes another point of view from *The Wizard of Oz*, and *Once Upon a Mattress* takes another point of view from *The Princess and the Pea*. What other stories deserve this treatment?**

Exploring the Causal Link Between Song Choice and the Performance of Professional Sports Teams

Michael Wade, York University, Toronto

Introduction and Conceptual Development

Interest in the performance of professional sports teams is widespread. Team owners want large returns on their investments, and those returns are generally greater when their teams perform well. The careers of players, general managers, and coaches often move in harmony with their teams' success. Sports fans, local businesses, municipalities, and other stakeholders also have a substantial interest in how local teams perform.

However, despite the importance of professional sports team performance to the broader economy, the research literature has been unable to accurately predict whether a team will win or lose on a given day. A number of factors have been identified that appear to affect game outcomes, including the quality of players and coaches, the extent and appropriateness of training, team cohesion, fan behavior, and home field advantage. Notwithstanding the fact that most of this information is widely available, researchers have been largely unsuccessful at accurately predicting the outcome of sporting events.

This research seeks to address this gap by studying the use and choice of music played through public address systems during games. Surprisingly, this important factor has not been extensively studied in the literature. In fact, the author could find no instances of music choice research in the context of sports team performance anywhere, despite the ubiquity of music in today's sporting arenas.

Theory clearly predicts that the choice of music can exert a significant influence on the outcome of games. Motivation theory suggests that players will heighten their level of play in response to rousing song choices (Maslow, 1943). Contagion theory suggests that athletes will seek to imitate the behavior contained within positively oriented music (Le Bon, 1895). Finally, cognitive dissonance theory proposes that participants in sporting activities will alter their conduct to close the gap between incompatible or conflicting environments (Festinger, 1957). Thus, when playing poorly, athletes will improve the quality of their play in response to songs that advocate positive behavior.

All these theories suggest that music choice in sporting arenas can affect the performance of home team players, and thus the eventual outcome of games. Thus,

Hypothesis: The choice of music in professional sports arenas will affect the performance of home teams.

Methodology

An empirical research methodology was employed in this research. There are 4 large professional sports leagues in North America: football (NFL), baseball (MLB), basketball (NBA), and ice hockey (NHL). Football was ruled out since there is no NFL team in the author's home city. Baseball was also excluded from the research due to a relatively small amount of music played during MLB games,[1] and the fact that baseball is played during the summer months, and thus games would unnecessarily impinge on the author's golfing schedule. Thus, the research included basketball and ice hockey. These 2 sports are ideal for the study as they are both played in indoor arenas and incorporate a large amount of music during games.

The researcher attempted to conduct the research by watching games on TV, but found that the confounding effect of commercial breaks and inane commentary adversely affected his ability to accurately determine song choices during games. Thus, the researcher obtained seasons tickets to the local NBA and NHL teams and attended every home game for both sports (including playoff games) for one season.[2] To reduce the effect of demand characteristics, the researcher brought along a friend, who also monitored the music played.

During games, the researcher and confederate kept careful track of music choices and scores, with particular attention to the final outcome of the game. Any music played during half time (for basketball games)

or period breaks (for ice hockey games) was excluded from the analysis due to the fact that players were not engaged in sporting activity during those times.

Results

The Table contains the most commonly heard songs during the games covered in this research, placed in order of frequency. Statistical analysis of the relationship between songs and outcomes for the home team is also reported in the Table. The results clearly show that song choice influences the outcome of games. Certain songs cause home teams to perform in a superior manner, as suggested by theory. For example, the song *We Are The Champions* by the British band Queen (whose lead guitarist, Brian May, earned a PhD in astrophysics) was consistently associated with winning performance. The author surmised that perhaps all songs by this band had a similar effect, a view that was supported by a second commonly heard Queen song, *Another One Bites The Dust*, that was also associated with home team wins. However, this line of reasoning was disproven when it was found that a third Queen song, *We Will Rock You*, showed no effect on game outcome.

Other songs that appeared to be harbingers of positive results were *Rock and Roll Part 2* (also referred to as the *Hey* song) by Gary Glitter, *Taking Care of Business* by Bachman Turner Overdrive, and *Can't Touch This* by MC Hammer (who had been a bat boy and more for the Oakland Athletics). This research strongly suggests that the management of professional sports teams should insist that these songs be played during games in their home arenas.

Most songs identified in the study exhibited no significant effect on game outcomes. Commonly heard songs like *Get Ready for This* by 2 Unlimited, and *Cotton Eyed Joe* by Rednex, did not lead to successful or unsuccessful conclusions.

Perhaps most importantly, the study found that certain songs had a negative effect on game results. These songs consistently led to losing performances from the home team. Despite the enduring popularity of the song *I Can't Get No Satisfaction* by the Rolling Stones, this music choice was clearly detrimental to home team performance. A similar negative effect was found for the Phil Collins song *There Must be Some Misunderstanding*. The evidence also suggested that this latter song led to abhorrent behavior from the home team's players, particular in ice hockey. Interestingly, a lack of music also seemed to be associated with poor home team performance.

The results of this study strongly support the hypothesis that song choice in sports arenas can significantly affect the outcome of games in professional sports. Thus, the paper finds robust support for motivation theory, contagion theory, and the theory of cognitive dissonance.

Discussion

The findings of this study contain clear implications for the management of professional sports teams. To positively influence the outcome, management should insist on playing a few key songs during games, while carefully avoiding others. If the team is playing poorly, for instance, this research suggests that a sensible course of action would be to play *We Are The Champions* or *Taking Care of Business*, perhaps multiple times, until team performance improves. This strategy is sure to please management, players, and fans alike.

While other factors such as player skill level, coaching ability, and so forth exert an effect on a team's performance, this study has shown clear evidence that choice of music is also an important factor in the determination of outcomes in professional sports. Future work will seek to extend the current research to other sports in additional markets. The author encourages collaboration with like-minded academics to expand and extend this important research program.

Notes

1 Preliminary analysis based on a small sample found no link between team performance and the song *Take Me Out to the Ball Game*.
2 The author would like to acknowledge the generous assistance of funding agencies in supporting this important research.

Song	Performer	Effect on home team performance		
		Win	Loss	No effect
Eye of the Tiger	Survivor			X
We Will Rock You	Queen			X
Get Ready for This	2 Unlimited			X
Rock and Roll Part 2	Gary Glitter	X		
Cotton Eyed Joe	Rednex			X
National Anthem				X
We Are The Champions	Queen	X		
Another One Bites The Dust	Queen	X		
Welcome to the Jungle	Guns n' Roses			X
Hells Bells	AC/DC			X
Can't Touch This	MC Hammer	X		
Taking Care of Business	BTO	X		
Who Let the Dogs Out	Baha Men			X
I Can't Get No Satisfaction	Rolling Stones		X	
There Must be Some Misunderstanding	Phil Collins		X	

Win signifies a positive significant correlation between the song and a team win ($p<0.05$)

Loss signifies a positive significant correlation between the song and a team loss ($p<0.05$)

No effect signifies no significant correlation between the song and the outcome of the game

References

Festinger, L. A., *Theory of Cognitive Dissonance,* Stanford University Press, Stanford, CA, 1957.

Le Bon, G., *The Crowd: A Study of the Popular Mind,* 1895.

Maslow, A. A preface to motivation theory. *Psychosomatic Med.,* 1943, 5, 85-92.

PROCEDURE : The student shall dissect a frog and examine its
internal organs.

Name of student . . . . E. MYRON FOGARTY

Labrotorie report on Cutting up a frog.

Plan. I will get a frog and a sharp nife, Which with I will cut opin the frog whilst he is alive. I will studdie carfulley his innerds, for the purpuss of gaining scientifik knolledge.

Part One. Mr. Higgens give me a frog, a nife, and a pan of stuff like jello. I grapt the frog by his hine legs and I beet his head on the edje of the lab bench so he wood not bight. When he was woosie and ready for scientific investigathun, I stuck thumtax in his foots and pinned him against the jello, with his bellie in frount. Part Two. Soon the frog begun ones agan to kick, and feering his eskape, I scientifikully rammed the shiv in his stumak. There was a stickie dark red floooid which ouzed out from whence I stabbed him. This was kind of fun so I druv the shiv in him agan. I noted that the more I stabbed him the more he kicked. Soon he begun to kick less and less each time I stabbed him and finnalley

he kicked not at all. I went to Mr. Higgins and got a new frog.
Part Three. I through what was left of the old frog in the wase
basket then put the new one in its place. I was determint to
learn more from this frog by condukting a more scientifik
experemint. I made too insessyuns akross his tummy and I peelt
the skin away so I coold see better. It was kinda silverry inside
so I cut deeper, and found his innerds, which I scientifikully
scoopt out with a spoon. Part Fore. I got a woodin handil
fork, upon which I put the frog. I turned on a bunsin
burnur and held the frog, who kicked not much too
terrible now, above it. Soon the smell of roastid frog
filled the labb and Mr. Higgens made me share him with
othur kids

Diagram of a frog which I learnt from this experiment.

eyebull
other
eyebull
lungs
glands
nose
bones
guts
tung
more guts
food
comes out
essofigus
hand
innerds
leg
Stumak

DIAGRAM OF
A FROG WHICH
I CUT.

The **Largest** Integer

Joel H. Spencer, Bell Telephone Laboratories

Mathematicians have long sought to discover the identity of the largest integer. Some have proclaimed that such a thing does not exist. This view, while possibly internally consistent, certainly cannot give a true model of the integers. For by symmetry, if there is a smallest integer, there must be a largest. Of course, this argument's premise may be, and has been, denied. However, without the principle of symmetry much of the work of the last 2 centuries would have to be discarded. This, we feel, is too great a price to pay. Our solution is given by the following:

Theorem: -1 is the largest integer.

Proof: List the integers
$$\ldots \; -4 \; -3 \; -2 \; -1 \qquad +1 \; +2 \; +3 \; +4 \; \ldots$$
You will note that nothing has been left out. The largest integer must have no successor, so it is clearly -1.

The Blinker Hypothesis

D. T. Arcieri, SUNY Farmingdale, & Christian Dietz
Arcieri, James Wilson Young Middle School

A very wise man once said, "There are 2 types of
people in this world: givers and takers." (1) This made
us reflect. Although it seems simplistic, this categorical
approach to defining fundamental human qualities does
have merit. One look at the myriad behaviors of
strangers in public, or friends and family in more private
settings, clearly underscores this premise. People do
behave many times in unmistakably generous or selfish
ways. Some open doors for you, but others cut in line in
front of you. Some bake cookies, while others eat them
all.

But can givers and takers be identified
unambiguously independent of overt behavior?
Certainly, complicated and subjective measures would
not serve this purpose. For example, having individuals
fill out long psychological questionnaires aimed at
differentiating the thoughtful from the thoughtless
would, no doubt, result in a skewed ratio favoring the
former over the latter. Therefore, a simple, objective
measure would be of more value with respect to the
accurate placement of individuals in one category or the
other.

Suggested in this study is a singular indicator: the
use of the blinker when driving. Automobile drivers, for
the most part, can be categorized as either those who
signal or those who do not. As signaling is the legal,
safe, and courteous behavior, we believe that those who
use their blinker are most likely givers. Since not
signaling is illegal, unsafe, and thoughtless, we take that
as sign that the driver is a taker. The purpose of this
study is to identify individuals as givers or takers, based

on whether they signal or not. More importantly, the cumulative data will tell us what percentage of the population are nice and thoughtful, as opposed to rude and selfish.

The authors sat on a street corner up the block from their house for half an hour. In that time they counted how many cars turned using a blinker, and how many did not.

The analyzed data indicated that 68% used their blinkers, and 32% did not.

Conclusion

If the driving population of our small town reflects society in general, then 32% of Americans, and possibly all of humanity, are takers. This is clearly a bummer and probably the fundamental reason the world is such a mess. Not that we can do anything about it. But it's good to know, anyway.

On a more practical level, identifying takers before they strike will be advantageous to givers. For instance, if the car behind you does not signal when turning into the movie theater parking lot, then don't sit in the theater near that driver. Obviously a taker, that person will talk during the movie, answer their ringing cell phone, and crinkle candy wrappers right behind you.

1. Personal Communication from Norman Dietz, D.T.'s father-in-law and Christian's grandfather.

☛ **How does this compare to your area? Can you make other observations that might have larger meaning?**

A System for the Simulation of Simultaneous Moves Between 2 Non-Colocational Players

Marisa Debowsky, New York University, &
Adrian Riskin, Mary Baldwin College

Abstract

We describe a new system for the simulation of simultaneous moves between non-colocational players. This has applications in the burgeoning Rock-Paper-Scissors-by-mail movement.

Introduction

Intensive study of competitive Rock-Paper-Scissors (RPS) [1] has revealed a crucial psychological aspect to the game as played by colocational participants. Expert players can spot their opponents' tells, discern predictable strategic patterns, intuit their opponents' connotational conceptions of the moves and thereby gain an edge, and so on. In these aspects, colocational RPS (CRPS) is much like poker, and, as in the game of poker, there are many potentially sound competitors who are unable to play at a high level due to shyness or other psychological conditions. Chess is another strategic game with a psychological component much like CRPS, and colocational play suffers from many of the same problems (or advantages, depending on one's point of view).

These problems are alleviated somewhat by the advent of online play, but to our minds the presence of an electronic intermediary spoils the primal purity of the games. Although this problem seems insoluble vis à vis poker at this time, it is elegantly and efficiently solved in the case of chess through by-mail play.

With this situation in mind, we set ourselves the task of designing an unbreakable system for by-mail RPS play that involves neither human nor electronic third-party mediation. We have designed such a system for 2 players. It is suitable for any game which requires simultaneous moves on the part of 2 players, so it works for RPS-25 as well as standard RPS [2]. Finally we describe a possible ranking system for federation play, such as those used in various chess federations. We close with an invitation to the reader to join our newly founded International Postal Rock-Paper-Scissors Federation (IPRPSF).

The 2-player system

The game of CRPS works like this: Alice and Bob say together "1, 2, 3, shoot!" One beat after saying the word "shoot!" they each make one of 3 signs with their hands[*]: a fist (rock), a flat hand with fingers and thumb extended and held together (paper), or index and middle fingers extended and slightly apart as in the "V-for-victory" sign (scissors). If the 2 signs thrown are the same, the round is declared a draw; otherwise, the winner is determined by the cyclic ranking: rock breaks scissors, scissors cut paper, paper covers rock.

For by-mail (non-colocational) play, Alice and Bob must communicate their moves to one another while each being certain that the other has made a move without knowledge of the opponent's move. We reject, on æsthetic grounds, any scheme requiring online communication.

One possible by-mail method would be for Alice and Bob to choose a particular date on which the move must be mailed, with the postmark verifying simultaneity. We reject this scheme for a number of reasons. First, it's too easy for things to go awry. A letter dropped in a mailbox or picked up by a letter-carrier may not end up postmarked on the intended day for reasons out of the control of Alice and Bob. It is possible to solve this problem by having the stamps hand-cancelled, but in a multi-round game this may involve a prohibitive amount of time (what with trips to the post office, waiting in line, etc.). Furthermore, this scheme does not insure against the bribery of a postal clerk, as unlikely an event as that may seem.

However, the most important reason we reject this scheme is social. One of the charms of by-mail chess is that the moves are contained in alternating letters, allowing the players to engage in a conversational correspondence as they play. We wanted to preserve this property in a NCRPS scheme. Finally, the scheme we came up with is much cooler than any of these other options.

The crucial observation is that with alternating rather than simultaneous play, only one player's move must be hidden. If Alice commits to her move but does not disclose it, then Bob is free to make his move in the clear, after which point Alice's move can be revealed. Each player has thrown a sign without knowledge of the other's move. Thus, we simply need a mechanism by which Alice can play without revealing her move to Bob, such that it can be disclosed later, ensuring that Bob can't peek before making his move, and Alice can't cheat and change her move after seeing Bob's.

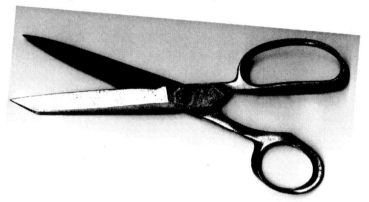

Our NCRPS scheme, therefore, requires a by-mail adaptation of a commitment scheme [3]. We simulate the hiding property of a commitment scheme using a sealed envelope, and we simulate the binding property using signatures across the envelope flaps. Consider the following 3-flow protocol: suppose Alice sends Bob a sealed envelope, signed across the flap. Bob receives the envelope, does not open it, and signs below Alice's signature before sending it back to her. Alice sees her own signature, so she knows that Bob has not replaced her original envelope with a fake one, thus ensuring that the contents of the envelope are still unknown to Bob (hiding). If she now mails it back to Bob, he will see his own signature and know that Alice has not replaced her original envelope with a fake one, thus ensuring that she has not changed the contents (binding).

Combining the 2 elements above, we present our scheme.

Protocol 1: *Alice writes her move on a piece of paper, seals it in an envelope, signs across the flap, and mails the envelope inside another envelope to Bob. Bob signs across the flap of the envelope containing Alice's move. Now Bob knows that as long as the envelope remains sealed Alice cannot change her move in response to his. Bob then returns the still-sealed envelope back to Alice with his move in the clear. Note that at this point Alice knows who has won the game. When she receives the envelope with the 2 signatures, she knows that Bob made his move without knowledge of hers. She then returns the sealed, signed envelope to Bob, who now knows that Alice has not changed her move in response to his. Bob now opens the sealed envelope and finds out who won the game.*

Traditionally, a match consists of a number of games played sequentially. While there is a certain amount of efficiency and no mathematically identifiable advantage in Alice enclosing her next move in her last mailing of the previous game, it is possible and even likely that there is a psychological advantage. Thus we choose to have Bob wait until Alice finishes off the first game by returning the doubly signed, still sealed envelope to him before sending her the first move of the second game, and so forth.

A Ranking system for federation play

A match consists of 10 games. Each player calculates the number of points a match is worth to him by subtracting the other player's score from his score. For instance, if Alice wins 7 games out of 10, then she gets 4 points on the match whereas Bob gets –4 points. After an official match, a player's ranking is calculated using the formula shown below, where R_n is the player's new ranking, R_o is the player's old ranking, P is the number of points the match was worth to the player, and R_o^A and R_o^B are Alice's and Bob's old rankings respectively.

$$R_n = \begin{cases} R_o + \left(\left[\dfrac{\left| R_o^A - R_o^B \right|}{5} \right] + \dfrac{1}{2} \right)(P+2) & \text{If lower ranked player wins or draws.} \\[2em] R_o + \dfrac{1}{2}P & \text{If higher ranked player wins.} \end{cases}$$

While this formula has not been tested extensively in federation play, it at least has the desirable property of rewarding or penalizing players more for upsets. Each new player, by the way, is awarded an initial score of 100.

☛ We invite interested readers, to whom the subtle pleasures of such an enterprise seems attractive, to submit for further information on joining the International Postal Rock Paper Scissors Federation, an eMail to the corresponding author: ariskin@mbc.edu

* There is a variant version where the signs are made at the same time that the word "shoot" is uttered, but we hold this to be mere pointless schismatism.

Bibliography

1. www.worldrps.com
2. www.umop.com/rps.htm
3. Giles Brassard, David Chaum, and Claude Crepeau: Minimum Disclosure Proofs of Knowledge. *Journal of Computer and System Sciences,* vol. 37, pp. 156-189, 1988.

Experiments Based on Commonly Held But
Seldom Tested Beliefs, Part II:

Unequal Gravitational Attraction Based on Jelly and Random Carpet Samples

Iva P. Aitchdee, Shipley Center for Immaterial
Science

It is a commonly held belief that the probability that
a piece of jelly toast will land jelly side down is
proportional to the expense of the carpet on the dining
room floor. I decided to test this theory using old carpet
samples from a carpet store, common white bread, and
grape jelly.

To eliminate variables, I used the same piece of
toast for the entire experiment, adding jelly to the toast
after each time it fell face down. I tried to maintain a
uniform jelly thickness of approximately 3 millimeters
above the surface of the bread throughout the
experiment.

I dropped the toast from the height of 5 feet,
allowing it to roll off of my finger tips just as I was about
to take a bite, thus accurately simulating the situation
of an accidental dropping. I dropped the toast 50 times
on each piece of carpet before my roommate came home.
As she could not understand the cause of sacrificing a
jar of her grape jelly in the name of Science, I had to
stop the experiment at that point or face a budget crisis.

Experimental data are shown in this table:

		Contrast with Face-Down	
Sample	Color	Grape Jelly Color	Percentage
1	Dark Blue	0.07	50%
2	Green	1.14	63%
3	White	8.96	87%
4	Tan	8.53	85%
5	Brown	3.38	60%
6	Mixed Vegetable	–3.85	52%

Color Contrast / Face-Down Percentage

Obviously the toast has a stronger tendency to fall face down on some of the carpets, but there is no correlation to price evident in the data!

To attempt to discover the pattern I arranged the data according to carpet thickness:

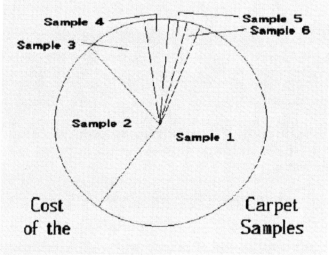

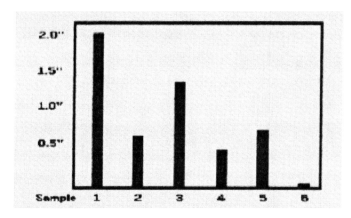

Thickness of the Carpet Samples

or carpet age:

1940	1950	1960	1970	1980	1990
	3	6	2 5	4	1

143

Timeline of Sample Ages

There was also little correlation with percent of synthetic material.

At this point, I realized that it is not the carpet price, but how much the color of the jelly clashes with the carpet that matters. The most expensive carpet sample tested was a deep purple-blue, which received a straight 50% of jelly-side-down landings. The white and tan carpets had the highest percentages of jelly-side-down landings, the green and brown less, and the cheap mixed veggie colored carpet had almost 50% again.

What we can conclude from this is that it doesn't matter so much the cost of the carpet that was ruined, but that if the carpet is ruined it has to be taken out and replaced regardless of its price. Toast has no motivation to land face-down on expensive carpet if that carpet will not be ruined in the process.

In light of this, homemakers are advised to coordinate their carpet colors with the sorts of things that might ruin them. If you use a lot of strawberry jam, a red or orange kitchen floor might be safest.

It is hypothesized that this can be applied to other areas of home maintenance. I would suggest grey carpets for rooms with fireplaces, brown carpets near the doors if your home is frequented by small children who like to play outside, and the ever popular mixed veggie institutional type carpet for the kitchen as it would survive the assault of a large lasagna casserole.

Without the effect of color contrast to increase the probability of a spill, ordinary probability dictates that only 50% of the time will your lasagna fall upside down, and even when it does, you will not have to replace your carpet.

Dr. Iva P. Aitchdee is the leading scientist at the Shipley Center for Immaterial Science and president of the Hoffman Foundation for Hypothetical Research. To learn more about her cutting edge scientific endeavors, visit www.rebeccajcarlson.com/review.html

Calculating Good Deeds

Joel Kirschbaum, PhD, Institute for Motivated Behavior

I decided to do a good deed daily.

Abstractly, I thought that this mission would be easy to achieve because I was currently contributing to several charities each month. I could attain my goal by either increments of a good deed a month, an almost infinitesimal rate, or immediately increase good deeds by one order of magnitude. I chose the latter, "Big Bang" benevolence.

But I stipulated rules: A single $31 gift couldn't translate to a month of good deeds. By extension I postulated that no one annual gift of $365.25 would satisfy my theoretical framework. Hypothetically it should be noted that one $3.66 charity check for the year may be quantifiable and statistically significant, but it's morally insignificant. I judged that to be equally outrageous as if I were to stuff a $1 bill into a different post-paid charity envelope each day.

Another premise was that 10 social "white lies", each approximately equivalent to 0.1 good deed, didn't equate to one good deed. I intended to do an

enumerable,

intact,

exact,

unequivocal,

person-to-person,

decent

deed daily.

An additional variable is the non-equivalence of good deeds. Therefore, in my calculus of pain and pleasure, I calculated that the boundaries of my bounty were a total of 30 minor and moderate good deeds, plus one additional massive good deed, to be mandatory monthly.

A further axiom I imposed was that holding a theater door open for many people only equaled one good deed — there is no mass-production of good deeds. It counted as but one good deed to have lent an unused room air conditioner to a neighboring family who had a broken house air conditioner, even though the duration was several days. Of course, while the heat exchanger fan rotated at 120 rpm, each day's total rotations did not translate to 172,800 good deeds — while each good deed equals a Jewish "mitzvah", there is no analogy to a Buddhist prayer wheel to generate multiple good deeds.

A single, linked good deed credited to several individuals occurred, however, when an unexpected box of chocolate-covered pretzels given to my wife was passed along by our daughter who needed a present to give to a frantic neighbor who had to have a suitable gift to bring immediately to a hospitalized teenager.

Here are the classes and categories of good deeds I found, as an aid in my actions and activities.

True Good Deed: Intuitively we know what an exact, traditional, good deed is, like giving unsolicited money to one's ex-wife's needy, distant, parents. A modern good deed would be freely donating either one's eggs or sperm to the infertile.

Good deeds are not necessarily limited by one's mortality. Donating one's organs after death to the tissue-deficient and deprived produces the good deed that keeps on amassing merit, even if anonymous.

Non-Good Deeds: Giving your young children an allowance which they must use to buy school lunches. Or sending your children to an expensive sleep-away camp for a month, or longer, if your primary, prioritized, personal intention is a vacation. Or giving any ethanol to an alcoholic.

Stealth (or Masquerading) Good Deeds: Situations where you hope to eventually reap a return. An example is using caller-ID to return a silent, "hang-up" telephone call to aid the caller about reaching the correct person and to give a long reminder about proper apology etiquette. Your true goal is to prevent another such interruption from the same person.

Another "hidden agenda" good deed is a donation to defray the deficit of a museum, symphony orchestra, or opera company that you especially enjoy — indirectly you benefit along with others.

A special case is giving a gargantuan gift with the goal of publicity, like having one's name on a building — a "Mount Rushmore Good Deed". Although this largesse shows how you tower among lesser mortals, such renown reaches and resonates in the realms of the rich, whose friendship eventually helps you increase your wealth to further allow you to make more massive good deeds. And thus the cycles spin. In addition, if you end up needing treatment in your donated hospital wing, you can be assured of the best care and attention possible.

Mischievous Good Deeds: For example, suppose on a lengthy flight you non-stop enlighten the morbidly-obese person stuffed into the adjacent cramped, coach seat, who is overflowing onto you, about overweight leading to diabetes, heart disease, high blood pressure, strokes and replacement joints. This is bestowing valuable advice, as well as subtle revenge for crowding you. And, even though both the individual's appearance in mirrors, and the advice of objective MDs, irrationally failed to inspire weight loss, there's an infinitesimal chance that your information eventually induces a shrinkage in the stranger's seat size to fit the usual sparse space.

Pseudo-Good Deeds: These are only imagined, or members of an empty set. An example is participating in a "roast" where your "comedic" comments consist of weird revelations about the recipient. Worse is a toast to the newly engaged or married, enumerating and detailing their adventures with strangers.

I don't always succeed in performing my daily good deed, but I'm improving. Which category does publishing this article fit into? Any and all advice to help me achieve my invariant goal would be a good deed credited to you.

Calculating More Good Deeds

Rod Embree, OD, Fort Wayne, Indiana

I read, with great interest, Joel Kirschbaum's "Calculating Good Deeds", above. Although I found it interesting and enlightening, it occurred to me that there were some additional considerations that Dr. Kirschbaum failed to mention.

First of all, what about credit for intent? For example, I might wish to do a good deed, but be unable to do so for reasons beyond my control. Factoring in the possibility that I am in denial and actually don't intend to do a good deed and am looking for an excuse, I estimate an intended good deed hindered would be worth 67.5% of an actual deed well done.

Furthermore, consider the eternal effect on our 'good deed quotient' when another person does a good deed in return. We might hope that helping an old lady across the street would accomplish the good deed for the day. But what if the old bag pulls out a $5 bill at the other end of the street in return for our help? Clearly this monetary offering has an effect on the deed that cannot be ignored. I would suggest that the net good deed has to be equal to the 'gross' or original deed MINUS any reciprocal deed. Of course, if you decline the money in this example, the reciprocal deed factor goes to zero, and if you knock out the old lady and steal her purse now that you realize she's loaded, the net good deed becomes a negative number.

I hope these few minor comments add slightly to the completeness of Kirschbaum's article. This, of course, is my true intent, which therefore leaves me with only 33.5% of a deed left to accomplish on this particular day.

MENDELEEV'S SONG

© 1997 Jeff Moran, "Dr. Chordate",
www.tranquility.net/~scimusic

How should I arrange these elements I know?
Alphabetical order seems a logical way to go.
I'll start out with aluminum and end up here with zinc;
That will look attractive and well-organized, I think.

    ... No, no, no, that will never do ...
There must be a better way, if I could only see,
Like value of contribution to our GDP.
The most important elements will be up at the top,
And when you reach the bottom ... you stop!

All 92 elements must go into this chart,
Arranged in rows and columns like a work of art.
I could have 4 rows of 23, or 23 rows of 4,
Or 5 rows of 10 and 6 rows of 7 ... who could ask for more?
    And I don't know just what I'd do if there were more than 92
    Elements to keep track of on my chart.
    From silver, gold, and platinum, on down to actinium,
    Everyone can tell them all apart.

    ... No, no, no. This will never do. It's too common, to ordinary.
This simply isn't innovative enough to get me a departmental chair ...
Wait! I've got it!
On the left I'll put the metal alkalis (whatever that may mean);
And on the right the noble gases would look mighty keen;
And all the other elements go somewhere in between,
As long as they are somewhere where they can be seen
    By the queen; better yet, by the dean.

    And I don't know just what I'd do if there were more than 92
    Elements to keep track of on my chart.
    From hydrogen and helium on down to uranium,
    Everyone can tell them all apart.
    My table will be stable, there'll be no decay.
    So I'd better write each letter without more delay ...

"How about neptunium?"   *Well, OK ...*
   Now I know there'll never be any more than 93
   Elements to keep track of on my chart.
   From hydrogen and helium on down to neptunium,
   Everyone can tell them all apart.

"How about plutonium?"   *OK, OK, plutonium, let's see ...*
   Now I know there won't be more elements than 94
   To keep track of on my periodic chart.
   From hydrogen and helium, on down to plutonium,
   Everyone can tell them all apart.

"How about ...?"   *Now, cut that out!*

The Montillation of Traxoline
Judy Lanier, Michigan State University

It is very important that you learn about traxoline. Traxoline is a new form of zionter. It is montilled in Ceristanna. The Ceristannians gristerlate large amounts of fevon and then bracter it to quasel traxoline. Traxoline may well be one of our most lukized snezlaus in the future because of our zionter lescelidge.

Directions: Answer the following questions in complete sentences.
1. What is traxoline?
2. Where is traxoline montilled?
3. How is traxoline quaselled?
4. Why is it important to know about traxoline?

☞ **Do you answer test questions *this* way, or do you *understand* what you're learning?**

Insect Rights

Dale Lowdermilk, Santa Barbara, California

With the furor over baboon organ transplants and animal experimentation, the National Organization Taunting Safety and Fairness Everywhere (NOT-SAFE) is proud to announce its support for the organizers and advocates of "animal rights". The abuse of laboratory animals, however, is a minor injustice when compared to the unspeakable atrocities which are committed each day, in every city, and even in our own homes.

Humanity is blindly and systematically attempting the equivalent of genocide on the INSECT world. Trillions (if not more) of innocent, peaceful, crawling creatures (including other arthropods, mollusks, annelids, etc.) are being slaughtered each day for such trivial "crimes" as:

- *living in wood* (termites)
- *stealing bread crumbs* (cockroaches)
- *congregating in garbage cans* (ants)
- *chewing on plants* (mites, locusts, worms) and
- *sleeping in hairballs or carpets* (fleas).

Just because a snail doesn't scream when stepped on, doesn't mean that it cannot feel pain.

These inadvertent "splatterings" are not limited to the USA but are occurring in South America with tiny beetles being crushed alive so their broken bodies and wings can be used as an ingredient in shellac products.

Obviously, some bloodthirsty bugs (like mosquitoes) MUST be killed, but even this could be done more mercifully if manufacturers simply added a few drops of anesthetic to each can of spray.

Contributions are now being accepted for the MedFly Preservation Fund, Montecito, California 93108.

If it's worth doing right, it's worth overdoing!

Physicist MacDonald's Farm

William C. Rands III

An elderly physicist named MacDonald happened to own a small farm. And on this farm he kept a large number of pigs. Old MacDonald was very interested in the value of his pigs, and especially the value residing at any specified point in the farm yard.

To measure his precious pig value, he first specified OINK, a scalar function of position referred to as O. This function could take one of 2 values, 1 or 0, depending on the presence or absence of a pig at that specified location. Since pigs distribute themselves haphazardly about the farm yard, he knew he would have here an OINK and there an OINK, but not everywhere an OINK.

He then defined 2 coefficients:
- E, the "extent" or volume of the pig at the specified position, and
- I, the "intensity" or density of the pig.

He observed that, as his pigs grew, their value went up as the square of their total mass, EI. Therefore, the pig value at any point in the farm yard could be readily seen to be proportional to:

$$(EI)^2 O$$
or EIEIO.

@n Exp°nent1@£ 1n¢®e@$e 1n P@$$w°rd ¢°mp£ex1t¥

Peter A. Stone, Columbia, SC

I work in a technical bureaucracy and thus experience most of the foibles of both realms — scientific and bureaucratic — and, worse, their synergistic interaction. I would like to record an example for ethnologists of the future, and to make a testable prediction. My data are observational, from prolonged intimate daily contact.

Computer passwords are necessary, of course: our armed guards can't keep out the kooks and at the same time keep an eye on all the employees. Cubicles to which many are consigned can't lock, scientific personnel can't be trusted to lock office doors in any case (ask security in any large federal facility), and there are at least a few of us with some trade secrets or private data regarding individuals. Potential vandalism is probably the bigger real worry. So the need for a password is acknowledged (as is, sadly, the explicit need to tell people with graduate degrees not to write those passwords on their computers or desks).

But here is the rub, the item of interest. We have bureaucrats in charge of password requirements, ones with no responsibilities for efficiency or output (being organizationally distant), and ones with enough savvy to know that there are a fair number of variables available via a keyboard. Nothing flashy, such as facial recognition scans through webcams (which I'm sure

153

have crossed their minds), but numerous characters and symbols and cases and fonts are readily available.

We started out with a simple password of some required reasonable length (details lost to history), with this minimum length being a single required variable. Other aspects could vary of course (use of letters, numbers, or keyboard symbols) but with no set requirement. This worked fine with no failures ever revealed, but after a long period the group in charge required 3 variables: upper AND lower case PLUS (number OR special character). Then in short order, one of our needed passwords (oh yes, we have several) required the previous, but number AND special character. The plotted curve so far looks to be exponential (unpublished data), safely assuming the last increase is not simply "noise".

Where do we go from here? I predict the sky is now the limit. They have the scent in their nostrils and they are given free rein in a bureaucracy, in part driven by the common delusion that what information **we** have must be **highly** important. And finally, they have a lot more room to work in. Let me give an example of the sort of requirements I predict will reign in just a few years, certainly within a decade, judged by the curve of increases to the present.

≥*10 characters* (iconic number with humans due to appendage geometry), **including:**

Upper AND lower case

At least one character each from at least 2 alphabets (e.g., Latin, Cyrillic, or Arabic)
At least one character in each of 2 typefaces or fonts (e.g., Helvetica, Batang, or Albertus Extra Bold)
At least one character or symbol each in **5** *of the following:*

　　Punctuation mark or linguistic operator or typographical symbol (e.g., !, ˜, or †)
　　Common special character (e.g., #, $, &)
　　Math function (operators, plus denoters such as ± and %)
　　Musical note
　　Playing card symbol
　　Emoticon (e.g., "smiley face") OR *Block element* (e.g., ▨)

Isotope in correct super- and subscript notation
(must exist, and counts as only one character: e.g., $^{238}_{92}U$, though preferably with the numbers correctly superimposed)

There will be many associated rules, not the least of which will be for letters that span different alphabet types (e.g., M in Latin, Greek, Cyrillic) or are used both as letters and math symbols (e.g., Σ). These complicating rules will not trouble the persons in charge in the least, but rather will give them satisfaction in providing another layer of security.

I give a few very simple minimal examples of passwords meeting these types of requirements that I predict for the not distant future, safely assuming that IT technocrats will remain independent and unrestrained. The internal orders of these examples are arranged as the requirements are listed above merely for clarity, as they could be changed as a matter of choice.

$$Ae\equiv\hat{}\ \Sigma♫♣☺\ ^4_2\textit{He}1$$

$$Qz\delta\check{}\equiv♪♥\ ^{237}_{93}Np\ \ 2$$

$$Zb℧‰\neq\leftrightarrow⅞\ ▐\ ^{134}_{51}\mathsf{Sb}3$$

Note (1) the element symbol, e.g., "He", cannot serve in the alphabet or case requirements, but **can** meet the second font requirement (as used in the examples), (2) the "Σ" is used here as a math operator, not the Greek letter per se, (3) the Greek letter "Ξ" is not the same as the math relational element "$\equiv$", (4) the "Sb" is not simply bolded but is a different typeface, and finally, (5) what may look like empty normal spaces are actually a white half of a block diagram and an "em" space, the latter a special character of sorts.

I predict that, soon thereafter, several colors will be required. Eventually, as more characters become available in standard software, these too will be added as options, for instance cuneiforms, runes, Egyptian hieroglyphics, Mayan pictographs, and Chinese ideographs (in the last case the multiple overstruck strokes, no matter now many, will count in sum as only one required character). Sadly, software also easily allows such requirements as "2 numbers, such that interaction via your secret personal math function (e.g.,

–Xe^Y) yields your secret personal constant", both of which will also need to be changed periodically. I am sure the clever person can easily envision additional attainable security along these same lines.

2 other obvious trends are (1) an increase in required frequency of changing of passwords, and (2) an increase in the needed number of passwords. With only 2 data points so far for the former, I cannot determine whether frequency increase will be linear or exponential. Data on the latter have been lost in time so this may simply be linear (in that no cluster of recent data points is obvious).

Finally, there has to be some upper limit here, to allow any other work to be conducted. The obvious relationship is:

$$\Omega = \Delta \times Z \times \Pi$$

where
Ω = an upper-limit constant (thank god!; note, this is not *THE* omega number of mathematical lore, or at least I don't think it is)
Δ = complexity,
Z = frequency (of change), and
Π = number (of passwords needed).

Psychosocial Aspects of Behaviour of Elevators

John T. Patcai, University of Toronto

Abstract

Background: Elevators are a prevalent feature of urban life. No studies have documented behaviour patterns of elevators as yet.

Method: An observational retrospective study: data in the author's organic neural net database were examined for common trends and features.

Results: Overwhelming evidence is presented that elevators are sentient, have predictable behaviour patterns, and have a social structure. Elevator / human interactions are designed by elevators to be deleterious to the physical and psychological health of humans.

Interpretation: As far as humans are concerned, elevators really know which buttons to push.

Introduction

Elevators have become an increasingly prevalent feature of public buildings in urban areas. In the experience of the author, hospitals are a very popular habitat for elevators in modern times. This was not always the case in the past. C. Darwin believed that elevators are the direct descendants of dumb-waiters (personal communication, séance, October 1999). The exact evolution of the elevator is not clear. Eventually, the Elevator Genome Project may shed some light on the evolutionary pathway that has brought us the modern elevator.

Methods

The author has examined the organic neural net database he has maintained in the urban Greater Toronto and Golden Horseshoe Area. This database was created with regular observations of approximately 40

157

elevators, and intermittent observations of hundreds of other elevators, over a period of up to 23 years.

Bias was avoided in that all elevators approached were entered in the study. No elevators were rejected for entry into the study for any reason whatsoever. However, some elevators were rejected for physical entry by the observer on arbitrary grounds.

The study was of a single blind construction, as the observer habitually closed his eyes while riding in an elevator.

The database was statistically analyzed using the Clinical Staff Tea Test, and only results approaching Moral Certainty with 95% Confidence Limits are reported below. The possibilities of Type I and Type II errors were reduced to negligible levels by using the spell-check feature.

Results

The following results were found to be valid, reproducible, and predictive.

1. Elevators always travel in packs for social reasons. Beware the unconventional elevator who defies the direction chosen by the pack.

2. A display of impatience by those awaiting an elevator causes elevators to arrive and travel more slowly. Signs of vertigo cause elevators to travel very quickly.

3. Elevators like to keep the doors peacefully open until a target rider is in the path of the doors. Then, the doors close rapidly. Elevators show no mercy to the old or disabled, and in fact these are the preferred targets. Fortunately, a rapid smack on the elevator's sensing apparatus will usually cause the elevator door to back off. Elevators are usually simple bullies. At heart, they are cowards.

4. Elevators have the ability to arrive at a position exactly even with the selected floor. Should the elevator realize that the rider is vulnerable, there will be a small discrepancy between the level of the elevator and the level of the floor. Riders must be vigilant to not trip and fall in this scenario.

5. The condescending nature of elevators can easily be seen if an elevator chooses to speak using a human voice.

6. Once the elevator has started to move, signs of agitated impatience in the rider will cause the elevator to skip the requested floor, or to reverse direction without warning.

7. People late for an important meeting, or sufferers of claustrophobia, need to take care to disguise the signs that the elevator observes. The elevator will usually stop between floors if it senses signs of extreme physiological stress in the rider.

8. Any sign of panic will result in the elevator rapidly dropping approximately one inch, resulting in an unexpected jolt. Any reaction from the rider will cause the elevator to repeat the behaviour, or exaggerate it.

9. Elevators may exhibit signs of illness behaviour. Some like the attention that repair people give them. Some become dependent on the care-giving repair people, or even addicted to the repairs. These elevators are constantly breaking down, and are easy to identify.

10. Elevators may become wantonly and uncontrollably destructive. They may drop several stories, or eat repair people for lunch. Fortunately, such episodes are very rare. However, there is no cure once an elevator has become a true rogue elevator. In that case, it has to be put down.

Interpretation

Elevators are sentient beings. They are able to discern many features of their riders, and change their behaviour accordingly. Elevators show social behaviour towards other elevators and show illness behaviour at times. Characteristics of elevators include bullying and cowardly behaviour. They seem to have a sense of humour, but it is not similar to the human sense of humour. Occasionally, one may meet an elevator with a pleasant personality, but that elevator usually doesn't fit into its peer group well. Rarely, one will find a destructive and dangerous elevator.

Some awareness of the environment by elevators has been shown in this article. Some cognitive processing by elevators is apparent. It is not known whether elevators have true self-awareness.

Replication of this work from other centres is awaited.

Futterman Scale of Sunrises

David K. Lynch, Topanga, California

1. Thick overcast, no colors, gradual brightening. Of no interest except to insomniacs, recent parolees, and lower life forms with little photosynthetic capability.

2. Thick overcast, occasional thin areas. No color except a dull blue lasting a few minutes. Gradual brightening. Occasional stirring in the underbrush. Damp breeze. Of interest only to roosters. No threat to Vampires.

3. Thin overcast, uniform grey. Color limited to dull blues with minor reds and yellows. Birds show some interest. Graveyards seem cheery. Vampires worried.

4. Thin overcast, some high clouds visible. Colors faded pinks and oranges, pleasant but unremarkable. More energetic birds take flight. First appearance of shadows and pale sunbeams.

5. Thin high clouds, scattered low clouds. Colors light pastels and airy, with occasional electric oranges accompanied by pink and yellow rays. Larks are heard and whole flights of wrens take to the skies. Much activity in the bushes. Distant violins can be heard. Breeze fragrant. Many promise to get up and enjoy the dawn more often.

6. Crisp air with cartoon animals in the clouds, amid a backdrop of cerulean blues and aquamarines. Sunbeams spring from every cloud. Positively aromatic breeze, warm and motherly. Grazing cows stop and look up. Birds harmonize with cats. People stop and stare, bouncy violin quartet heard even indoors. Bambi and Thumper seen. Most sensitive people feel faint and must sit down.

7. Clouds billow with multicolor pulsating fringe lace, sunbeams dance and sparkle. First cherubs appear around loftier buildups. Great sense of well-being. Golden horns sound from above and drown out full orchestra. Fragrance of ambrosia quite pronounced.

Bluebirds and robins fly sweeping formations, and nocturnal bats remain aloft. Cats and dogs dance jigs. Great nations agree to discard weapons. Most people swoon if looking directly into sunrise. Artists drop like flies. Even southern Californians admit that it's far out (sic).

8. Clouds resemble Venus de Milo, David, Pieta. Colors almost indescribable as entire celestial vault throbs, sending out wave after wave of shimmering color. Cherubs everywhere. Cupid seen darting between great trumpets. Music swells and approaches climactic crescendo, drowning out the pandemonium in the bushes, and audible in deepest coal mines. Few people remain standing, and many pass out, even those indoors and not directly illuminated. Breeze overpoweringly fragrant, and carries all known species of birds aloft. Only 1 out of 10 poets still conscious. Continental drift and evolution momentarily stop. Photographs impossible.

9. Details sketchy. Only 8 such sunrises reported and only 2 verified. Eye witnesses limited to hardened criminals who saw sunrise reflected from bulletproof glass. #9 sunset suspected of causing dinosaur extinctions.

10. Theoretically possible. Probably fatal to most life forms, even those with little photosynthetic capability.

Construction with Copper Discs

Mitch Fincher
All photos from his www.CoinStacking.com

I held a scrap of paper in my hand with the address scribbled on it. "Floor 42, apartment 12, Tokyo Towers South". I took a deep breath and knocked. Being a freelance journalist, I never know how these cold calls will turn out. I had come to Tokyo last year on a hot story and stayed. Now my ideas for stories were drying up and I had bills to pay. One good story for an in-flight magazine would pay last month's rent.

An elderly Japanese man with white hair answered the door and said, "You are the reporter?"
"Yes, and you are Dr. Fujitsu of Tokyo University?"
"Yes, yes. Come in." he said quickly in excellent English.
I entered his small, cramped apartment and noticed the framed photographs of various buildings clinging to the walls. Dr. Fujitsu had designed 4 skyscrapers in Japan, one in Paris and 2 in L.A. Before he retired to teaching, he was *the* in-demand architect of Japan.
He motioned me over to his dining table by a window.
I had to wade through the clutter, stepping over a scale model of a traditional Japanese pagoda on the way. The view of the skyline was breathtaking. I knew now why a man of his stature would live in this cramped apartment.
"Tea?" he asked.
I took the tea to be polite and slurped the best I could.
"You are here about my amazing new idea?" he asked.

I assured him I was, although a little fuzzy on what the idea was. A friend had given me his name.

"My grand idea came to me during an earthquake", Dr. Fujitsu said eagerly. "During the quake I was surfing the web and viewing a site on coin stacking – a craze that started here in Japan, of course, and has swept over the world. The coin structures are built using only the weight of the coins to cantilever coins over open space; no adhesive is used. That's when it hit me – I could design earthquake-proof structures from huge copper discs! I was shocked that no one had thought of this before."

"For example," Dr. Fujitsu said showing me the following photo, "Bridges built like these would be earthquake-proof because the coins just slide over each other during a quake."

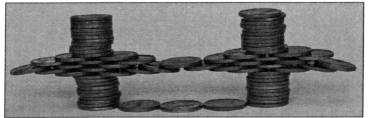

Photo by Sarah Fincher

"So you want to scale these structures up to cross rivers?" I asked hesitantly.

"Yes, yes, of course." he said, somewhat irritated that I was so dense.

"And this picture shows a design that can be used for long spans" he continued unfazed.

Photo by Mike Peden, Dripping Springs, Texas

"I am going to change the way buildings are made and make a mint in the process. For example, this tower will be a high-rise apartment building. Each flat will be its own self-contained prefab hollow disc with plumbing routed above and below on flexible tubes to the neighbors – windows all around the edges of course. With prefab construction we have a solution to the Japanese housing shortage! We can stack these earthquake-proof structures to the sky" Dr. Fujitsu babbled.

"Our nuclear power plants will be much safer with cooling towers made like these" he said, again sliding a glossy photo my way.

Photo by Ben Kiefer

"More tea?" he asked.

I declined and looked at my watch. This hot tip I'd received from a friend about a genius architect seemed a little tarnished now.

"No longer will earthquakes take a toll on our country." he said.

"Well, how exactly does this work?" I asked.

"Only the weight of the upper units holds the bottom ones in place. The friction between the coins is the coefficient of friction times the weight of the discs above. The upper ones will sway, dissipating the energy of the earthquake with their friction. The structures will be flexible like bamboo, so they will survive the earthquakes. Afterwards, hydraulic jacks will be used to nudge the units back to their correct place. This is much cheaper than repairing the twisted steel frames of traditional structures."

164

He continued, "And with dome structures we can have sporting events hosted inside multi-story earthquake buildings."

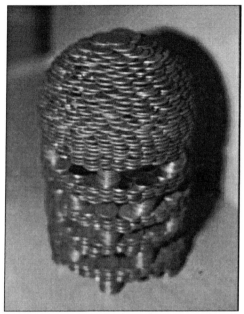

Photo by Mike Peden, Dripping Springs, Texas

Dr. Fujitsu continued to talk and showed more photos, saying that he could design office towers like these.

Photo by Steinar, Norway

Photo by Jeff Preshing, Montreal, Quebec

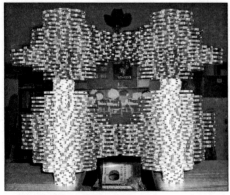

Photo © Rick Einhorm

Photo by Todd Kettering, Austin, Texas

"Of course" Dr. Fujitsu said, "the whole world will go to my new method – it's only common sense."

I was sure by now that the good doctor was a few fries short of a happy meal. Not wanting to leave hurriedly, I asked what he wanted to do with the rest of his life.

"I have a yen to leave my mark on the world." he said frankly.

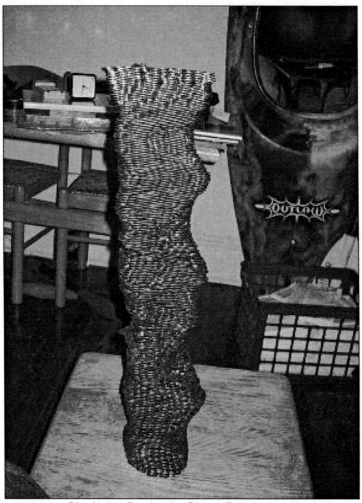

Photo by Peder A. Olsen, Tonsberg, Norway

I thanked him for his time, put my tea cup down
and excused myself.

Leaving the building, I entered the crowded street.
The ground began to shake. I almost yelled in fright.
The girders in the building above me groaned and a
window shattered, but the crowd just swayed gently with
the Earth, unaffected, and continued on their way
unconcerned.

I thought to myself, "Maybe Dr. Fujitsu is right and
Japanese architecture is ready for change."

☛Cheaper to build with pennies or yen than with quarters.

Conduction Heat Transfer Between a Little Girl's Bottom and a Metal Playground Slide

K. Schulgasser & O. Igra

Introduction

In recent years several workers have investigated both theoretically and experimentally the conductive heat transfer from sitting animals to the cold ground or to cold floors. Some of this work is summarized in reference 1. Among other creatures, sheep, cows, swine, ducks, and Brooklyn-fed cockroaches have been considered. The present work goes far beyond these piddling efforts in that it considers:

1. A non-steady-state situation with interface heat generation due to friction.

2. The extent to which experiments performed on animals can be used to infer human response (and vice versa).

3. Wear and tear.

4. Changing of the conductive properties of the interface layer without changing those of the substrate, that is, a sheep cannot change its fleece but a little girl can change her pants.

5. A situation wherein the possibility exists of communicating with the subject; thus something of the human-engineering aspect of the problem can be learned.

Experimental Procedure

A rather small sample of 4-year-old female humans, in good health, were placed on a standard 35° playground slide (specifications may be found in reference 2). The experiments were carried out in controlled environments and the slides were preconditioned in these environments for 24 hours: at 0°F for one series of experiments (chill bottom conditions), and at 130°F for a second series (hot seat conditions). The subjects were given a controlled push, and various precise measurements were made. Runs involving torn clothing, crying, side spills, and vomiting were disqualified. To induce maximum cooperation, subjects were fed to satiety immediately preceding each test run. As might be expected, repeatability was high. Since it was found that any kind of lubricant significantly affects test results, the slide was thoroughly dried and degassed prior to each test run. For comparison, similar experiments were performed with sheep; incidentally, they seemed to enjoy their new experience.

Results and Conclusions

In all cases the experiments showed and the theory confirmed that when the slide was cold, heat initially flowed from the subject to the slide. When the slide was hot the reverse was true. The subjects' subjective opinion, expressed in in-depth interviews, invariably agreed with these findings. The sheep were not interviewed. Although not included as one of the goals of this research, it should be noted that for the cold runs, the subjects invariably preferred flannel panties, while for hot runs, the "no panties" option was preferred.

Some of our quantitative results are reported in the graphs. It is seen there that the theory and experiments agree remarkably well. At first glance some of the sheep data would seem to imply appreciable scatter. This was at first attributed to the apparent unwillingness of the sheep to cooperate. During their descent they fidgeted and sometimes even bounded vertically upward. However, the theory shows that this is no random, disturbing phenomenon. The theory follows precisely the *apparently* erratic jumps in the curve. It should here be mentioned that it was considered prudent to include in the theoretical formulation a number of arbitrary parameters, 17 in all. Because of this fortuitous near-perfect agreement between theory and experiment, there is no need to show error bounds for the data, or even to estimate experimental error; there apparently was none.

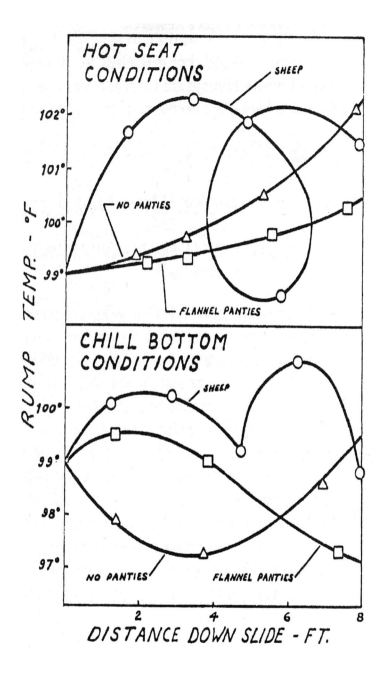

Rump temperature as a function of distance down slide: data points and theory (solid lines).

Suggestions for Future Research

This work should be extended to determine whether the above findings are valid also for males and for adults. Furthermore, the effects of varying the roughness of the slide surface and of various lubricants should be investigated. Another important parameter, not as yet investigated, is how gravitational changes (which directly influence sliding time) affect the considered heat transfer. Since our laboratory is located only some 600 feet above sea level, it is highly recommended that the above described experiments be repeated at some appropriate high altitude Himalayan location, in spite of the difficulties which might arise in connections with the "hot runs". A proposal to cover this planned research has been jointly submitted by us and by the University of Cat Balou to the Sherpa Academy for the Advancement of Science. Let us hope that the cash is forthcoming. Our bags are packed.

Acknowledgement

The authors wish to thank Yael and Hadas for their cooperation. They are brave little girls.

References

1. R. M. Gatenby, "Conduction of heat from sheep to ground", *Agricultural Meteorology* 18, pp 387-400 (1977).
2. MIL-SPEC 76R-32L-18R – Slide, Playground, All-weather, Steel, Uncoated (Unauthorized revised inversion 1974).

The Search for Even-Valued Prime Numbers

Howard I. Amols & Lawrence E. Reinstein

The search for even-valued prime numbers has received little attention in recent years. Most mathematicians have taken the easy way out, concentrating their efforts on discovering and cataloging the much more commonly occurring *odd*-valued prime numbers. Early investigators, however, were intrigued with the possible existence of even primes.

Although failing to find an example, the great 16th-Century Portuguese mathematician Rinaldi Jesus de Medeiros published an elegant, albeit little-known, treatise on even prime numbers. Unfortunately, this work was never translated and, in fact, was lost in 1537.

Further progress was made by an amateur Albanian mathematician, Balek Organian, who demonstrated that even primes could be identified with a large electronic computer. Unfortunately, Organian presented his hypothesis in 1846, before computers were available. The search for even primes effectively died with Organian in 1853.

We, however, have reopened the search for these exciting numbers.

Our search was based on 2 fundamental theorems. First, the number 2 is both even, and prime! This fact alone defeats all attempts to prove the non-existence of even primes. Surely, if there's one even prime number, aren't there others? Second, numerous examples exist of so-called type 1 sequences (primes which differ by 2 or 4):

3, 5, 7
3, 7, 11 ... etc.

Further, type 2 sequences exist with prime numbers differing by 3 or 5:

$$-1, 2, 5$$
and $$-3, 2, 7.$$

Since numerous type 1 sequences exist, the possibility arises that additional type 2 sequences might also exist, and the second number of such a sequence must be an even prime!

Our search for even primes has been centered on series of "near" type 2 sequences – so-called Daled sequences. The first Daled sequence is:

1) 61, 64, 67
2) 101, 104, 107
3) 131, 134, 137
4) 151, 154, 157

That is, the Daled sequence consists of 4 "near" type 2 sequences. In all 4 sequences, consecutive numbers differ by 3 (a requirement of type sequences), the first and third terms of each sequence are prime, and the second term of each sequence is even (though not prime). What defines the above as a Daled sequence is the relationship between consecutive sequences. Namely, the first number $j_i(1)$ of each sequence (i) may be represented as:

$$j_i(1) = j_{(i-1)}(1) + 60 - 10i$$

where the subscripts refer to sequence number (i = 2-4).

Consider now the prime factors of the second number of each sequence:

$j_1(2) = 64 = 2 \times 2 \times 2 \times 2 \times 2 \times 2$ (6 prime factors)
$j_2(2) = 104 = 2 \times 2 \times 2 \times 13$ (4 prime factors)
$j_3(2) = 134 = 2 \times 67$ (2 prime factors)
$j_4(2) = 154 = 2 \times 7 \times 11$ (3 prime factors)

Note the pattern for i = 1 to 3, namely, the number of prime factors of $j_i(2)$ decreases from 6, to 4, to 2. Unfortunately, for i = 4, the number of prime factors again increases to 3.

However, Fermat's Unknown Theorem[1] says that other Daled sequences may exist for which the number of factors of $j_i(2)$ continues to decrease. Thus, a $j_4(2)$ may exist with only one prime factor. Such a number must be an even prime.

A computer program written to test all Daled series has been running continuously on a Cray/ola computer, but no even primes have yet been found. The failure to find an even prime number at this early stage of the calculation, however, is hardly surprising, for several reasons. First, since the number of odd primes is infinite, then the separation between even primes must conversely approach infinity. Since the number 2 is known to be prime, the next even prime might in fact be infinitely large! Secondly, number conservation theory[2] states that:

    1) There are an infinite number of primes

    2) There are an infinite number of numbers

    3) Every other number is even.

Clearly, conservation theory requires that at least some even numbers be prime, but again, they might be infinitely large.

Hence we are continuing our computerized searches for such numbers.

Although it is now obvious that even primes do in fact exist, we have become increasingly suspicious that such numbers may not be integers and in fact may even be imaginary.[3] Revisions in our computer codes will allow us to test this possibility in future calculations.

Finally, we note that some numbers – already proven to be prime, and thought to be odd, may in fact be even!

This work was supported by USDA Division of Prime (Beef).

    1. The appellation "Unknown" for this theorem developed as a consequence of the fact that no reference of any kind is made to this theorem in any of Fermat's work.

    2. *Big Bird Looks at Numbers,* by Bird, B. (Hought-Muffin).

    3. Consider, for example, the imaginary number 4i, which is even, but which cannot be expressed as the product of 2 real prime integers. Is 4i an even prime?

Unembargoed Report of Department of Defense Strategy #183,000: The Military's Latest, Most Devastating Secret Weapon

Lynne V. McFarland & Marc McFarland

Introduction

The *New York Times* published an article detailing military plans for the increasing use of robots and robotics in warfare. (1) While interesting, this shifts focus away from psy-ops' (psychological operations) deepest, most secret weapon in development:

Birds of prey. And by birds of prey, we can mean only one thing:

The Shell Parakeet. Also known as the budgerigar in some parts of the world.

AKA: The budgie.

Your enemy's worst enemy. (Figure 1)

The Problem

One of the main problems in any military operation is flushing out an enemy from hiding places in areas the enemy knows best. These can be in the back-alley mazes of ancient cities, in the deepest recesses of the darkest jungles, in the 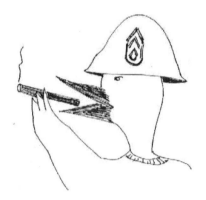 plunging shafts and treacherous caverns of the world's most rugged and remote mountain ranges, far beneath the shifting dirt and sands of the most unforgiving deserts, etc. (2)

The chief problem is, the enemy knows the territory. You don't. How to flush him out? Do you risk personnel and equipment in ferreting out your quarry? Or is there a way of minimizing exposure of your troops and assets while maximizing opportunity to force your hidden enemy to surrender?

Not all conventional psychological methods of war have been effective in our current conflicts. (2, 3) This may be in part because one study of these methods, listed in PubMed, is an article written, and we quote: "in Undetermined Language." (3) Understandably, no abstract is provided, and a trip to South America is needed to obtain the original article (no such trip was included in our original grant application).

A new, innovative, cost-effective, low-risk approach is urgently needed that is effective and yet conforms with humanitarian guidelines. (4)

The Solution

Psy-ops has discovered that loud, unpleasant noises and pop music blasted at top volume drives an enemy first to distraction, then to despair, then to complete demoralization and, finally, to willing, unconditional surrender. For example, US forces blasted loud rock music at the Papal Nuncio's residence to drive Panamanian strongman Manuel Noriega out of hiding and into the arms of his captors during "Operation Just Cause" in 1990. (5)

While effective, this works only when you know the exact location of your enemy. Unfortunately, this specific information is not always available. You may know the area, but the precise location of your quarry remains a mystery. Besides, if your enemy grew up with rock 'n' roll blasting away through the headphones of his iPod or Discman, he may very well be inured to this. Heck, he'd probably even *like* Metallica blared at top volume at 3 AM.

Psy-ops personnel are, at this very moment, teaching budgies to talk, to turn them into weapons that will bring any hidden enemy, no matter how well-concealed, out of hiding, crying openly and profusely, pleading — even demanding — to be taken into custody. Anything, so long as he gets away from the budgies.

Psy-ops personnel are teaching 100,000 budgies to squawk, repeatedly and in unison, "I'm a pretty bird!" These meticulously-trained budgies will be released into areas where an enemy is hidden. The theory is simple, but effective: No single individual, nor groups of armed insurrectionists, no matter how devoted they are to their cause, can withstand the constant verbal bombardment of 100,000 budgies yelling "I'm a pretty bird!" for any appreciable amount of time. In beta testing, the strongest-willed subject endured 37 seconds of the "I'm a pretty bird!" onslaught before running into the open, surrendering, and kissing his captors gratefully as they took him away (Figure 2).

Risk Assessment

The risks to troops and non-budgie assets is minimal. Pilot studies showed that, as long as troops were supplied with earplugs, the effect was limited to the target population. Non-human, non-budgie populations (cats, dogs, and goldfish) were unaffected. Risks to the budgies were found to be variable. This strategy may be potentially hazardous if the enemy is well-armed, but psy-ops theorizes that any more than 30 seconds of "I'm a pretty bird!" renders most subjects incapable of discharging a weapon. Or of doing practically anything else, for that matter.

Costs

This strategy should be very cost-effective. Markov economic models indicate the only significant cost variable would be birdseed, but as the total cost of birdseed would be $0.025/bird/day, the total cost would be $7,500 for a 3-day operation. Compared to feeding human troops, food costs for 100,000 budgies are, well, chickenfeed. Perhaps a contractor, such as Haliburton, could provide the government with a reduced bulk rate. Wait! What are we thinking? ...

Conforms to Tenets of the Geneva Conventions?

Will this strategy conform to humanitarian guidelines? (4). Our legal department says, "Guess so."

Preliminary Conclusions

In addition to flushing out an enemy, if successful, this might even make a mean "Movie-of-the-Week". If not, we're pretty sure we can fund our next study by getting it on "TV's Funniest Videos".

References

1. I'd quote the paper, but it's lining my birdcage and my ill-tempered parrot has a huge beak.
2. Beltran LJ. Principles and application of psychological method in the war propaganda. *Med Cir Guerra*. 1954; 16:3-11.
3. Rydeheard DE, Mycroft A, Peto J. Propaganda wars. *Nature*. 1982; 296:700.
4. Slim H. The continuing metamorphosis of the humanitarian practitioner: some new colours for an endangered chameleon. *Disasters*. 1995; 19:110-26.
5. Noriega source: www.gwu.edu/~nsarchiv/nsa/DOCUMENT/950206.htm

Parents' and Researchers' Views of Children's Invisible Imaginary Companions

Richard H. Passman & Espen Klausen, University of Wisconsin-Milwaukee

[THIS SPACE IS INTENTIONALLY LEFT BLANK.]

Authors' Notes

The authors thank the pretend companions, Harvey the Rabbit, Winnie the Pooh, and Calvin's Hobbes for their assistance in ghost-writing this article. Their suggestions were of unimaginable help.

In accord with the American Psychological Association's standards for the ethical treatment of participants, no imaginary companions were hurt in the preparation of this article.

Empirical Measurement of m&ms Contained in a Standard Bottom-Mouth Erlenmeyer Klein Flask and Comparison to Theoretical Models

Bernard Y. Tao, Dept. of Useless Information in an Effort to Win a Free Klein Flask, A Large Midwestern University Somewhere in the Soybean/Corn Belt

Abstract

While m&ms and Klein bottle geometry ostensibly have little technical or economic value in common, research was undertaken to establish theoretical and empirical values for the number of m&ms that could be contained in the Erlenmeyer Klein flask (EKF) shown. This research was performed solely for the purpose of obtaining such a flask absolutely free, either with or without m&ms.
(www.kleinbottle.com/m&ms_in_a_klein_bottle.htm)

4 researchers at a major Midwestern university independently developed methodologies to estimate the number of m&ms contained in the aforementioned vessel, using data provided by the original problem proposer. 3 theoretical models were developed to estimate values, using a standard Erlenmeyer flask as a model for the Klein analog. An empirical measurement was used to confirm the validity of all models, along with a statistical analysis of m&m size.

Empirical measurement found that the EKF contains 549 ± 3 m&ms. One theoretical model produced values within 1% of this number, but others ranged as high as 14% difference. However, there may be complicating factors in the method, producing the higher error value.

Introduction

Klein bottles are 2 dimensional topological structures that exist in 3 dimensions, having zero volume. However, they have been used for a variety of entertaining purposes, as noted in (1). Given the highly multi-disciplinary nature of this problem, it was thought that a single approach would not be as successful in developing useful solutions. Therefore, we attempted to develop solutions using physical chemical, mathematical, sociological, and engineering approaches. The objective of this work was to obtain a completely free Erlenmeyer Klein flask as noted in (4). However, if available, a completely free Klein Stein of similar volume (5) would be preferable.

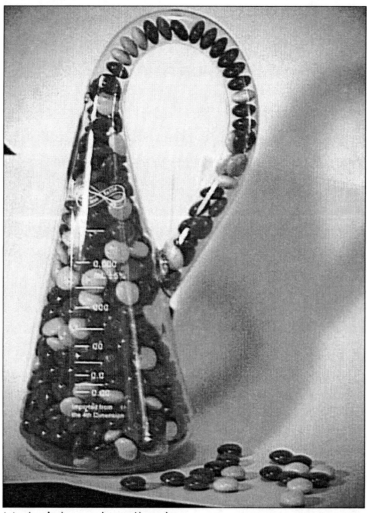

Materials and methods

The Erlenmeyer flask used to approximate the Klein analog was obtained from Fisher Scientific (Pittsburgh, PA), model 4980, 500 mL Pyrex. 2 packages of m&ms were obtained from a local grocery store (m&m/Mars,

Hackettstown, NJ, net wt. 14.0 oz, milk chocolate). All calculations were performed on a Sharp Scientific calculator (model EL-5100S, Sharp Corp., Korea) using pre-installed algorithms for transcendental functions. 15 cm ruler used to measure m&m oblate spheroid radii was from Davis Liquid Crystals, Inc. (San Leandro, CA). All other materials used were of reagent grade or better.

Method 1: Physical Chemistry estimation

Assume m&ms are oblate spheroids with a major axis radius of 0.6 cm (a) and a minor axis radius of 0.3 cm (b). The volume of an oblate spheroid is $(4/3)\pi\, a^2b$. Assuming hexagonal close packing (hcp), the void volume % of the packaging is $(4/3)\pi\, a^2b/(\pi/1.2092)$ $a^2(4)b = 0.4031$, or 40.31% of the available volume is composed of m&ms (2, 3).

Assuming the Erlenmeyer Klein flask is basically a right angle cone, we can use its dimensions of 240 mm x 100 mm (h x D). The volume of a cone $[(\pi/3)*(D/2)2*h]$ with these dimensions is 628.32 cm^3.

This means the effective volume filled with m&m mass must be approximately 628.32*0.4031 = 253.26 cm^3. Dividing this value by the volume of a single m&m (0.4524 cm^3), gives the number of m&m in the bottle as 559.8.

Method 2: Mathematical estimation

Alternatively, consider the specified volume of the flask. Given that the original volume of the Erlenmeyer flask is 500 cm^3, this must be corrected to account for the volume of the neck, which is approximately 110 ml, so the total volume would be 610 cm^3.

Multiplying this value by the void volume of hcp oblate spheroids (0.4031) gives a volume of 245.89 cm^3 of m&ms. Dividing this volume by the volume of a single m&m (0.45239 cm^3) gives the total number of m&ms as 543.53.

Method 3: Liberal arts estimation

Fact: m&ms are mainly composed of chocolate.

Fact: Chocolate is recognized to have a transcendental power on bipedal mammals, particularly female humans.

Fact: The best known transcendental numbers are e and π.

Fact: Normal human female mammals have 20 digits, which are used to eat m&ms.

Fact: "Erlenmeyer Klein flask" has 20 letters.

Armed with this knowledge, we can estimate the number of m&ms in the flask by multiplying the number of bipedal mammalian digits used to ingest m&ms by the transcendental number e to the power of π and add the number of letters in Erlenmeyer Klein flask to obtain:

$20*e^\pi + 20 = 482.81$

This demonstrates that without an extensive knowledge of mathematics or physical chemistry, one can also simply estimate the number of m&ms in an Erlenmeyer Klein flask. However, as normal, this also demonstrates that a liberal arts degree is essentially worthless in technical computations.

Method 4: Engineering estimation

2 Bags of m&ms were purchased and poured slowly into a 500 mL Erlenmeyer flask from the lab. This was followed by equilibrating the system using gentle agitation and addition of m&ms to fill to the lip of the vessel. Subsequently, the m&ms were poured out into large plastic weighing dishes and manually counted. Ignoring the broken and chipped ones, which were used for personal metabolic studies, the number obtained was 549 ± 3.

Results and Discussion

Results of the 4 methods are summarized in Table 1.

Table 1. Raw data
Method 1, Chemist: 559.8
Method 2, Mathematician: 543.5
Method 3, Liberal Arts: 482.8
Method 4, Engineer: 549 ± 3

There is inherent error in any of these calculations or estimates, given that the angles and details of elongation and extension of the neck of the flask to form the Klein nexus are unknown (see photo). Unfortunately, accounting for these errors is not possible without additional data on the geometrical issues.

It was found that although the assumption of an oblate spheroid for a single m&m is quite reasonable within the precision of the available instrumentation (my eye and a 15 cm ruler marked in 1 mm increments), the radii of the long and short axes of an m&m were precisely 0.6 cm and 0.3 cm, respectively. However, extensive statistical analysis of the contents of 2 14.0 oz bags of m&ms, approximately 1100 pieces (aren't graduate students terrific!), demonstrated that the mean mass of m&ms is 0.911725 g, with a standard error of 0.03174 g. Ignoring the density changes between the candy coating and the chocolate interior, this converts to a potential error of 3.48% in the volumetric calculation of a single m&m. Using this variation, the numbers previously obtained by methods 1 and 2 must be adjusted to yield the following corrected values (see Table 2):

Table 2. Corrected values

Method 1, Chemist: 559.8 ± 19.5
Method 2, Mathematician: 543.5 ± 18.9
Method 3, Liberal Arts: 482.8
Method 4, Engineer: 549 ± 3

There are several obvious conclusions that can be developed from these results. First, the engineer's method (method 4) is highly accurate under the current situation. It employs the fewest number of assumptions. However, the method employed is wholly empirical and does not account for the theoretical nature of m&m structure, the complexities of m&m packing within the vessel, or the statistical issues involving population variation among the sample. Therefore, while valuable for the purposes of this contest, it cannot be extrapolated to other shapes, sizes, or situations.

The chemist's method (method 1) clearly has the highest absolute error value, nearly 20 m&ms. Additionally, the assumption of hexagonal close packing (hcp) is clearly in error, given the graphical evidence shown on the photo. The m&ms in the nexus clearly demonstrate body centered cubic packing (bcc), not hcp.

Method 2 yields results with a similar statistical error value, but gives a value that is approximately 3.0% lower than method 1. However, given the statistical error value, it is clearly close to the actual value as found by counting. This is probably due to the greater accuracy involved with measuring the additional volume of the flask neck, vs. the assumption of a conical shape, as in method 1. Further improvement of method 1 might involve using a truncated cone assumption combined with a short right cylinder analysis to obtain an improved estimate of the additional volume of the neck of the flask.

Method 3 clearly has significant shortcomings vs. the other methods. This is not surprising, given the highly heuristic nature of the methodology employed. The value obtained is nearly 14% different from the values obtained by other methods, although the methodology employed is highly appealing and very simple. It does not account for structural geometry, physical chemistry, or statistical variation, and has very little sound theoretical mathematical basis.

Additionally, the veracity of the experimentalist may be in question. This question was raised due to the discovery that following the experimental procedures, approximately 25% of the original mass of m&ms provided to the researcher were absent. She noted in her lab book that this may have involved "cold fusion, high m&m vapor pressures, or other unspecified errors". Additionally, a lab assistant noted a mysterious brown smudge on the researcher's lips, although this information is purely anecdotal.

Further research in this area may involve extension of the theoretical models developed herein to other geometries of Klein bottles, notably ellipsoidal, cylindrical, and modified spiral, provided suitable research funding or free vessels could be obtained.

Editor's Note

By actual count, the flask held 547 m&ms.

References

1. Stoll, Clifford, Acme Klein Bottle, www.kleinbottle.com, Oakland, CA.
2. Castellan, G. W., "Structures of Solids and Liquids", Chap. 26, in *Physical Chemistry,* 2nd ed., Addison-Wesley, 1971, pp. 633-637.
3. Hoerl, A. E., "Plane Geometric Figures with Straight Boundaries", *Perry's Chemical Engineers' Handbook,* 6th ed., McGraw Hill, 1984, p. 2-11
4. www.kleinbottle.com/m&ms_in_a_klein_bottle.htm
5. www.kleinbottle.com/drinking_mug_klein_bottle.htm

R.E.V.E.L.A.T.I.O.N.?

excerpted with permission from the book by Milt Pupique.

www.PublishAmerica.com, 2002, 1-59129-233-6.

The title is the acronym for "**R**eality **E**xpressed by **V**irtually **E**xplicit and **L**urid **A**cronyms of a **T**itillatingly **I**nsightful and **O**ffensive **N**ature." Actually, a lot of these acronyms aren't luridly offensive, they're cute and clever. Here's a sampler:

Four Letter Word: **W**ord of an **O**bjectionable or **R**eprehensible **K**ind
That Which is Lost Forever: **M**isplaced **A**rticles **I**rretrievably **L**ost
How the Television Networks Try to Present the News: **F**actual **U**pdated **N**ews
Horoscope: **C**ertifiably **R**eproducible, **A**ccurate, **C**elestial **K**nowledge **P**ertaining to the **O**rder of **T**he **S**tars
Overbearing Boss: **M**odel of **E**fficient **M**anagement **O**rganization
Junior Executives: **C**ompany-**O**riented **G**overning **S**taff
A Real Power Tie: **N**otably **O**stentatious **O**bject **S**uitable for the **E**xecutive
A Bird from Brooklyn: **B**rooklyn **O**rnithologically **I**nvolved **D**eterminant
The Source of All the Energy in the Universe: **C**osmically **A**ssociated **F**ormative **F**orce for the **E**ternal **I**nduction of **N**ascent **E**nergy
Cow: **M**ammarily **O**perant **O**rganism
Dead Cow: **N**ecrotically **O**riented **M**ammarily **O**perant **O**rganism
The Cafeteria Dining Experience: **E**ating at a **C**ulinary-**O**riented **L**icensed **I**nstitution
Dog Food: **C**anine **A**limentary **T**rophic **S**ubstance
Road Kill: **T**raffic and **H**ighway-associated **U**nintended **M**echanical **P**rey
Hospital Emergency Room: **W**ard for the **A**lleviation of **I**llness and **T**rauma
Cannibalism: **I**t is a **T**ruly **T**errible **A**bomination of the **S**atanic **T**ype, **E**xemplified by **S**ociopathic and **L**unatic **I**ndividuals **K**illing and **E**ventually **C**onsuming **H**umans

in an **I**nstitution of the **C**ruelest **K**ind of **E**vil **N**ourishment

Whom Modern-Day Psychiatry Has Proven to be Responsible for One's Actions: **O**ne's **T**rue, **H**olistically and **E**xistentially **R**eal **S**elf

Quick Snack for Kindergarten Kids: **C**ompletely and **R**eadily **A**vailable, **Y**outh-**O**riented **N**utritive **S**ubstance

Very Shy: timorously introverted and of a modestly-inclined demeanor

Hard of Hearing: the **W**eak of **H**earing and **A**uditorily **T**roubled (???)

Permissive Classroom Management: **L**iberally-**O**riented **R**egime and **D**iscipline, **O**ffering **F**reedom and **F**ull **L**iberty to **I**ntelligent and **E**ducable **S**tudents

Schools: **S**chools and **K**indred **I**nstitutions of **N**ominally **N**urturing **E**ducational **R**esources **I**nvolving **S**cholastic **M**aturation

Large Universities: **M**ajor **I**nstitutions **L**inked to **L**earning and **S**tudy

Employment Opportunities for Recently Graduated PhDs: **P**rofessional **H**ack **D**river

Academic Science Research: **H**igh-tech **U**ndertaking of **B**asic **R**esearch of an **I**nstitutionalized **S**ort

The Rulers of the Future: **N**ominal **E**lite of a **R**apidly **D**eveloping **S**ociety

The Essence of a Truly Original Thought: **N**ovel **O**riginal **T**hought

The Road Always Traveled: **T**he **R**ace's **U**niversal **T**rail

The Sad Truth is That We Could Easily Be Replaced at Work By: **M**echanically **O**riented **R**obotically **O**perative **N**on-human **S**ystems.

The Era of Modern Technology: **G**olden **A**ge for the **D**evelopments of **G**reat **E**lectronic and **T**echnological **S**uper-Innovations

Men's Toilet Seat: **U**rinally **P**re-set

Women's Toilet Seat: **N**atural and **O**ptimal for **T**ransfer-of-**U**rine **P**rocess

The Essence of Give and Take: **T**ransactionally-**A**ssociated **K**ey **E**xchange

Deep-Space Hand Salutes

I'm the Captain. Don't ever forget it.

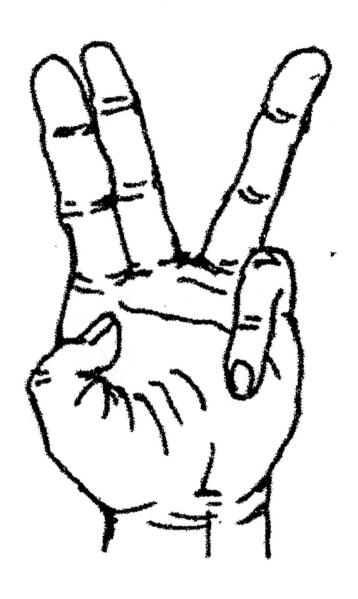

Our starboard warp drive thruster just fell off.

We're in reverse! Have the engineers check
wire winding directions on all electric motors.

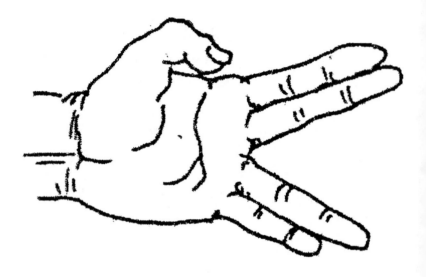

Silhouette show in the rec room at 1900 hours.

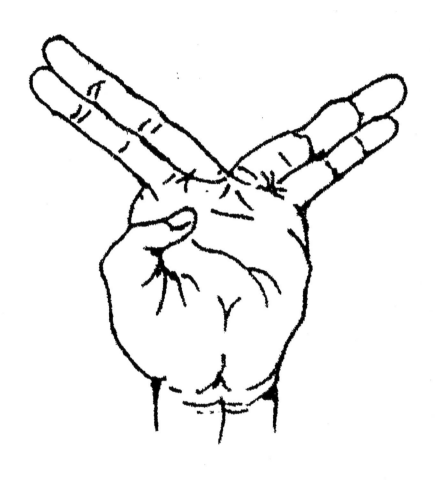

Captain Kirk threw his back out again.

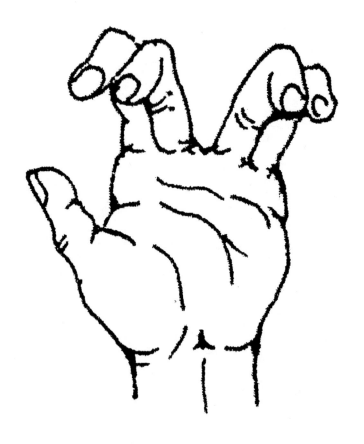

Who farted?

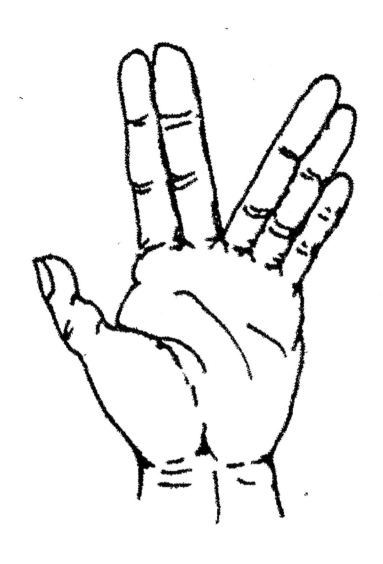

General Quarters! We may have aliens in our
midst!

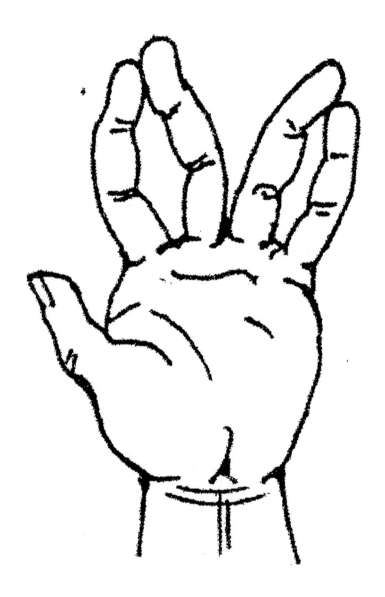

May you eat space worm holes for breakfast.

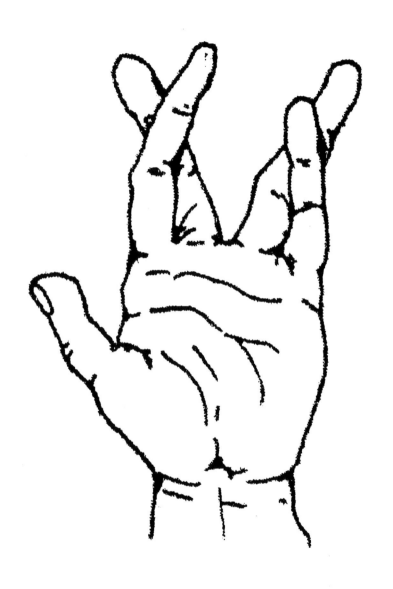

OK, Go talk to the Klingons if you must. But don't tell them any classified information.

One, over easy, with hash browns.

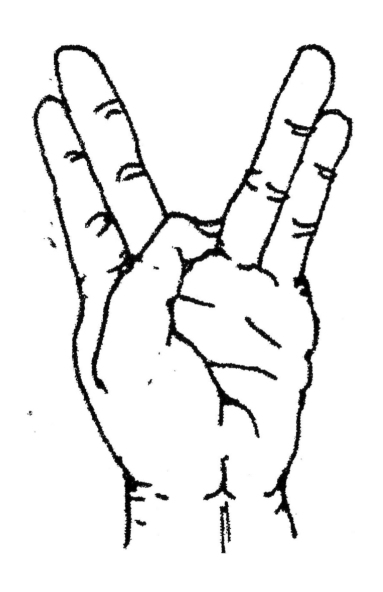

Request permission to go pee.

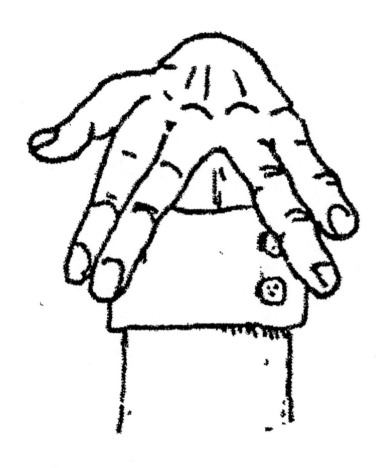

Stow the greeting. You're on Report.

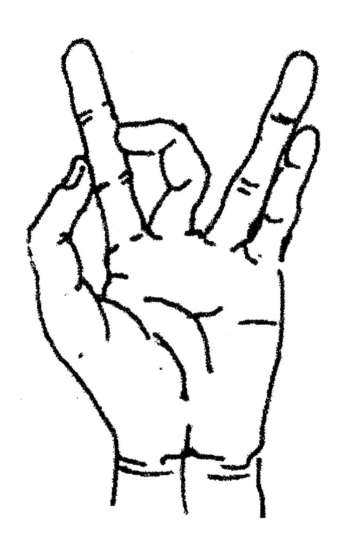

You're under arrest.

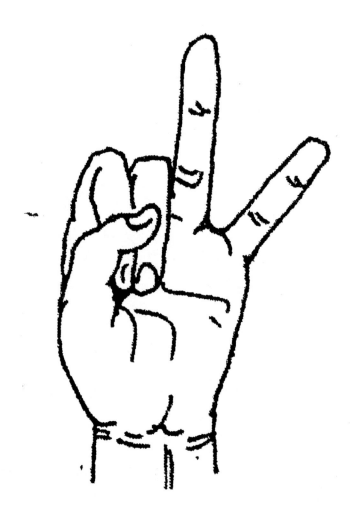

Welcome aboard, Second Lieutenant.

Nixing Ixodes

Michael R. Stevens, PharmD

Selected Highlights from *1,001 Ways to Kill a Deer Tick*

The blacklegged tick, *Ixodes scapularis*, is a
common vector of the Lyme disease spirochete, *Borrelia
burgdorferi*, as well as human babesiosis *(Babesia
microti)*, human granulocytic ehrlichiosis (HGE), and
other associated maladies. The highest concentration of
the species occurs in the northeastern United States.
Correlations between high concentrations of both deer
ticks (blood sucking) and political liberals (tax sucking)
in this area remain largely unexplored.

The origins of *I. scapularis* remain hidden in the
mists of time. Current entomological thinking is that
the species originated in Hell, as reflected in the species'
previous name, *Ixodes dammini.*

As a person living with Lyme
Disease (PLWLD), I have
developed an interest in ridding
the earth of this particular
scourge. While a planet-wide
dusting with DDT could possibly
be effective, I have found the
amount of paperwork necessary
to obtain proper authorization
precludes this approach. I have,
therefore, taken the time-
honored approach of "one tick at
a time."

Obviously, one person alone cannot accomplish this
monumental task. I have decided to involve the world's
population in this effort through my book, *1,001 Ways
to Kill a Deer Tick*. If each of the world's 6,477,744,000
inhabitants uses the information in the book, we could
eliminate 6,484,221,744,000 (± 13) ticks after just one
cycle.

Listed below are selected examples of ways in which
the common person can dispose of any and all deer ticks
encountered. The only additional purchase necessary
for women is duct tape. No purchase is necessary for
men, as we all already have copious amounts of this

"grey wonder" tucked away throughout our homes, boats, and automobiles.

Number 005: The Saturn V Rocket

Upon capture of the tick, firmly duct-tape it to the uppermost (third) stage of a Saturn V rocket (the first 2 stages may return your tick to Earth). If you do not already have such a rocket, check with your local Army Surplus Store as they may have one or more just taking up space. Often times, these stores are more than happy to give these obsolete rockets away to make room for cruise missiles and the like.

Once the tick is affixed to the rocket, light the fuse. Stand well away.

Number 051: Area 51

Perhaps the only sure-fire method for removal of deer ticks from this reality plane. The tick is simply duct-taped to a common postcard. The postcard is addressed:

U.S. Government Secret Installation
Alien Autopsy Unit
051 Roswell Avenue
Groom Lake, NV 00000-0000

The tick may not truly disappear, but – for all intents and purposes – it will have never existed.

Number 289: Wishing Well

Upon capture of the tick, tape the evil creature to a standard bowling ball. Locate an abandoned, yet open well. These can often be found in back yards frequented by small children. Simply drop the tick/ball complex down the well after first checking to confirm that no small child will impede the ball's journey into the dark abyss below. If you cannot tell if there is such an obstruction, you may utilize a Collie as a detector. While Collies cannot identify all children who may be in the well, they can determine if there are any named "Timmy" lodged within.

Number 289-A: We Wish You a Marianas

In this variation of the "Wishing Well" method above, simply stow away on any boat leaving a coastal city (usually San Francisco in the USA) for a voyage to the Challenger Deep in the Mariana Trench, 200 miles (320 kilometers) southwest of Guam. Under cover of darkness, the bowling ball is "inadvertently" dropped overboard. The tick/bowling ball composite will reside forever at a depth of over 35,840 feet (10,924 meters) below the surface. As an added bonus, the tick will eventually be subducted into the earth's mantle as the Pacific tectonic plate is forced to dive beneath the Mariana plate.

Number 575: Sock-It-To-'Em!

Perhaps the "cleanest" way to dispose of the little bugger, the captured tick is duct-taped securely to a freshly-washed sock and sent though the dryer. Take particular care to ensure that the sock selected is one of an intact pair, as unpaired socks never disappear. Both socks are placed into the dryer and the cycle is begun. At the end of the cycle, the little-known Heizenberg "Certainty" Principle (Sock variant) ensures that 0.08% of all such sock dryings will result in one of the pair transitioning to an alternate universe.

If, due to chance, the remaining sock is the one containing the tick, simply reattach the tick to another sock in an intact pair. Lather, rinse, and repeat until the preferred sock disappears.

Number 893: Call Guido and Forget About It

The author knows this guy who can make just about anything disappear for a price.

Contact me and I will see if he is free for your "job".

Number 893-A: I Know This Other Guy

The author can also put you in touch with David Copperfield, who can also make things disappear. However, this approach may require you to fly to New York to duct-tape your tick to the Statue of Liberty.

These are just a few of the methods I have devised which are contained in my upcoming book. Look for *1,001 Ways to Kill a Deer Tick* at all finer bookstores and grocery checkout stands ... once I find a publisher.

Photo: Scott Bauer, Agricultural Research Service

On Using Snakes to Increase Student Appreciation for Mathematics

Lawrence M. Lesser, PhD, University of Texas at El Paso

Studies show that math makes some students recoil. Researchers led by Drs. Vy Perr and Fang Shui conjectured they could help students appreciate math by using it to help them shed an even greater phobia for them: snakes.

This juxtaposition is not as strange as it sounds, for the very word reptile (from 'repetitive tiling') is a shape with the property that it tiles a larger version of itself, using identical copies of itself. A simple example is a square, because 4 copies of it tile a larger square. A triangle also is a reptile, since 4 copies of it tile a larger version of that triangle. Similarity of shape invokes the concept of scale, and scales are something all snakes have!

The researchers gathered a group of students and took them to a snake lab to experience a sequence of activities, and we report the results (all hypothesis tests were one-tailed, of course).

First, to illustrate how mathematics is the science of patterns, students were tested (using live snakes, for real-world experience) on how well they distinguished the sequence "red, yellow, black, yellow, red, yellow, black, yellow" of venomous coral snakes from "red, black, yellow, black, red, black, yellow, black" of harmless milksnakes.

Then, there was an activity involving measuring and estimation, as the students studied snake striking range (y) as a function of snake length (x). Students arrived at the result $y = .51x$ after several trials (and errors).

Then, students observed from what geometric formations snakes could not (e.g., a line) or could (e.g., a curl; and just imagine the curl of a whole vector field of snakes!) strike. This part of the research was initially confusing because in baseball, the operational definition of "strike" is a miss.

After learning that while snakes are sensitive to vibration, they cannot really "hear" (with the possible exception of choral snakes), students raised the classic philosophers' question "if a rhombusback, I mean diamondback, rattlesnake alone in the desert shakes its rattle, does it make a sound?"

Cobras were found to be the snake most resistant to using complex numbers, as they seemed to spit venom towards any i's, so students learned to use the hood (the lab's hood, not the cobra's). Students also discovered that they greatly preferred lengthy penny-tossing probability experiments to handling copperheads.

Studying snakes' forked tongues was chaotic, but helped students understand bifurcation theory. There was a dramatic moment of discovery at one point, when a long snake was constricting itself in circles around one student. He insistently yelled out "pi!" (Or was he saying python? I guess we'll never know.)

In the end, the surviving students reported a statistically significant increase in their attitude towards mathematics relative to snakes and acknowledged that there was actually a lot of practical mathematics

involved in studying snakes — especially the adders, who are said to thrive near fallen trees because adders "need logs to multiply"! Hey, what's that hissing sound?

Gustatory Preferences for Marmite® versus Vegemite® Amongst Americans

Christopher J. Gill & Paul Bolton, Boston University

Abstract

To resolve one of the burning culinary questions of our era, we conducted a single-blinded taste test to determine whether participants preferred the taste of buttered sandwiches topped with Marmite or Vegemite. To avoid bias due to allegiance to Commonwealth Nations that favour one yeast extract (YE) over the other, we conducted this experiment amongst a group of American 'volunteers', none of whom had any prior exposure to YE sandwiches. YE sandwiches were contrasted on the basis of average scores on a 4 domain, 3-point scale; as a binary outcome; and on the basis of a semi-qualitative selection of descriptives. Our results suggest that Americans were unable to tell the difference between the 2 YEs, though dissatisfaction with the gustatory experience was nearly universal. A small subset of Americans greatly enjoyed both YEs, however. Further research is needed to understand in what ways the sensory functions of this gifted subset differ.

Introduction

Within the Commonwealth, citizens of Australia and Great Britain include many devoted fans of yeast extract (YE) spreads. These are marketed under the names 'Vegemite' and 'Marmite' respectively. YEs are an excellent source of B-complex vitamins, making them essential in the prevention of Pellagra, Beri-Beri, Wernicke Korsakoff's psychosis, and pernicious anemia. YEs trace their roots to beer making, originally being the tenacious residue coating the inside of a brewer's kettle after fermentation. While one might think that this pungent glue-like compound would be best thrown away without further consideration, this highly ignorant point of view emerges from a lack of respect for the resourcefulness that underlies many of the choice delicacies that make Commonwealth cuisine the envy of the world. Verily, the introduction and mainstream acceptance of both Marmite and Vegemite constitute one of the greatest marketing coups of the modern era!

Today, yeast extract sandwiches served on buttered bread (or as a tempting seasoning for soups) stand proudly alongside other Commonwealth delicacies such as 'Bangers and Mash', 'Eel pie with peas', 'Spotted dick', and 'Hole in the toad'!

Predictably, there has been considerable debate as to which YE derivative is the more delicious and satisfying. Marmite is 'yeast extract' whereas Vegemite is 'concentrated yeast extract'. Both are salty and savory, and, when applied in judicious proportions, become a tempting treat for the discerning palate. While it is intuitively obvious that Marmite should be superior, coming as it does from Old Blighty (Mother England), a surprisingly vocal minority comprised of those from 'Down Under' dispute this common sense viewpoint, insisting that they 'pity the fool' who prefers Marmite[1].

Several scientific conferences have been held to settle this debate, the most recent attended by members of the Australian Rugger Outback's league, and booster supporters of the Manchester United football club. However, despite some poignant verbal sparring and sharp exchange of fisticuffs (and some beer bottles), the ultimate superiority of Marmite remained disputed. At least one modern theorist[2] has pointed out that, coming from a nation founded by criminals and political exiles, Australians could hardly be counted on to pass qualified judgment on issues of high culture. Witness the popularity from Brisbane to Perth of such monstrosities as 'Waltzing Matilda', 'Crocodile Dundee', and the persistent and bewildering popularity in Australia of ABBA songs. In contrast, Australian culinary 'cognoscenti' stridently hold that the British palate is an inevitable consequence of living under the constant threat of rain with a nagging sense of French-envy[3].

Accordingly, scientific delegates from England and Australia[4] agreed that the ideal solution would be to have the issue settled in America amongst Americans. Moreover, Canada, as with many other current or former Commonwealth nations, already enjoys the health benefits and luxury of Marmite at the breakfast table, whereas Marmite has, quite tragically, been nearly unobtainable in the continental US. For this reason, we elected to conduct a blinded controlled side-by-side comparative taste test among a group of American public health professionals at the Boston University

School of Public Health – none of whom had any prior exposure to YE spreads.

Methods

Open-faced sandwiches were prepared according to the following protocol. We selected 10mm-thick slices of white bread purchased at a commercial supermarket in the US as the 'sandwich base'. The bread was used fresh, visibly free of mould, and un-toasted. To each slice, we applied precisely one tablespoon of pre-salted, room temperature, pre-conditioned fresh creamery butter, spread evenly over the bread surface to within a 2mm margin of the crusts on all 4 sides using a standardized spreading utensil. The researchers then applied either Marmite (CJG – representative of Great Britain) or Vegemite (PB – representative of Australia) according to their best estimate of the ideal ratio of butter to YE. This was also deemed necessary to avoid jealous contamination of the experiment by intentional over-application of the spread by the opposing Commonwealth Nation's spread's representative (CJG and PB respectively), recognizing that the potent flavour of Marmite/Vegemite causes even devoted fans of yeast extract to weep when it is ingested in excessive volumes.

Sandwiches were then cut into 6 roughly equal, bite-sized mouthwatering morsels. One square of bread with Marmite was left intentionally unused due to the presence of a large air hole, but was instead consumed by Dr Gill (with great pleasure).

Participants were asked to consume sample 'A' first, then 'B', and then to complete the questionnaire regarding the experience. Items included a question asking simply whether the subject preferred the sandwich prepared with YE 'A' or 'B', a series of semi-quantitative items relating to the 'appearance', 'smell', 'taste', and 'after-flavour' of each spread, and a semi-qualitative grid data matrix for both spreads in which they checked off any number of pre-defined categories of descriptives for 'A' / 'B' that they deemed appropriate. Participants were not informed as to the identity of either YE until the completion of the study.

Our sample size was determined empirically on the basis of the number of colleagues fortunate enough to be present in our office suite on Friday, August 5, 2005, when we arbitrarily decided to conduct this experiment.

The IRB was not consulted ahead of time, though generally food taste tests are considered exempt from IRB oversight.

We used χ square to compare the proportions of participants favouring 'A' over 'B', and paired *t*-tests to compare the mean subjective scores on each of the 4 descriptive domains: taste, appearance, smell, and after-flavour. Qualitative responses were tabulated histographically. All data entry and statistical analysis was conducted using SPSS version 11.0.

Study results

Quantitative analyses: 16 people were approached for participation in the survey. 2 were excluded as both came from Commonwealth countries and were already familiar with Marmite and had generally favourable impressions (one person was from Ghana, the other from Hong Kong). 14 people completed the taste testing, of whom one was originally from India but had lived in the US for many years and had no prior experience with or opinion of YE spreads.

Overall, 9 preferred Vegemite while 5 preferred Marmite. However, the difference was not statistically significant (χ square 1.143, p = 0.3). In the descriptive evaluations, there were no significant differences observed between Marmite or Vegemite, though surprisingly the category scores trended towards somewhat negative impressions (p value for trend was NS) (see Figure 1).

Interestingly, though more subjects expressed a preference for the taste of Vegemite in the quantitative analysis, the descriptives trended towards a consistently more negative impression of Vegemite than Marmite (p value for trend = NS). Several participants offered marginal notes in addition to those pre-coded on the case report form. One person commented that they preferred sample A (Marmite), but that both A and B were 'repulsive'. Another noted that the smell of Marmite was 'like a dead rat'. This person, perhaps coincidentally, expressed a preference for Vegemite. It should be noted that 'dead rat' is not often found on the restaurant menus of most Commonwealth nations, nor a frequent item in most chain supermarkets. Another elaborated that the after-flavour of Marmite was akin to 'gym socks', whereas the after-flavour of Vegemite was reminiscent of 'toe jam'. There was no information provided by this subject to infer how they became familiar with these particular flavours to the degree that they could confidently draw such subtle distinctions.

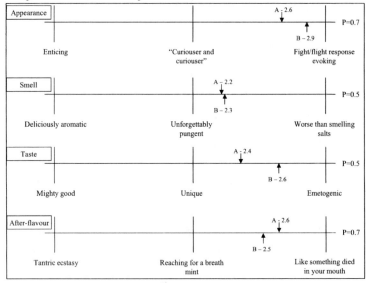

Figure 1

Qualitative impressions of Marmite vs.

Vegemite: As can be seen from Figure 2, the majority of our American subjects offered somewhat unfavourable impressions of both Marmite and Vegemite. Note, these categories are not linearly distributed in their estimation of the relative merits of the savoury spreads in question. Moreover, respondents were free to select as many of the categories as they felt were appropriate in their response to each YE spread. No respondent mixed favourable with unfavourable responses within their appraisal of either Marmite or Vegemite. However, the impressions of several respondents were strikingly divergent between the 2. One respondent felt that Vegemite was like 'concentrated pond scum', whereas Marmite was 'like a velvet climax'. In contrast, another felt Marmite to be akin to 'concentrated pond scum', while describing Vegemite as 'mother's milk'. Yet a third felt that Marmite evoked 'Natalie Cole's *Unforgettable*', with Vegemite reminiscent of 'arthropod jelly on toast'. We should note however that opinions differ sharply about the merits of Ms. Cole's song, with some feeling it to be the musical analog of internment in a Turkish prison, as depicted in the movie *Midnight Express*. Others find her music not inoffensive, perhaps on the same par as the collected works of Olivia Newton John.

Discussion

This controlled, blinded study showed that our American test subjects were not only unable to distinguish between Marmite and Vegemite reliably, but inexplicably found both spreads highly repellent. However, it is noteworthy that in the qualitative analysis, a subset of approximately 5% (1 person) expressed great enjoyment of both. Clearly further research is needed to identify in what ways this more rarefied subset differs from the common herd of Yankee yeast extract objectors (YYEOs).

While we cannot exclude the possibility of sampling error in this small pilot study, this finding at least raises the hypothesis that Americans have no sense about the finer aspects of high culture – a sentiment voiced throughout the Commonwealth. Some researchers have hypothesized that Americans' general inability to speak intelligible English may constitute a risk factor for their failure to recognize delicious gourmet delicacies. Circumstantial evidence in support of this theory is Americans' near-criminal mispronunciation of 'Worcestershire sauce'[5] as an explanatory factor in the near absence of proper fish and chip shops in the US, and the complete absence of British salad cream on US restaurant tables.

Others scholars identify the lingering resentment of Americans towards England in the face of overwhelming evidence of British cultural superiority. Notable examples of this include *Monty Python's Flying Circus*, Mick Jagger, the British Royal Family, and the actor Hugh Grant. This resentment emerges from a deep seated – and probably justified – concern that the outcome of the 18th-Century war for US independence was not appropriately ratified in the court of international law of that day, a consequence of which is that America technically remains a dependent colony of Great Britain. The implications in regards to past arrears for the tea tax alone are staggering. Moreover, this would require a complete reinterpretation of the credit due for the outcome of the Second World War. In any case, it is highly ironic that such views about Marmite and Vegemite would be held in a nation peopled by devotees of peanut butter – a particularly nasty and wholly unnecessary contrivance.

We were unable to find a definition for the 'toe jam', mentioned by one of our subjects, in Stedman's medical dictionary, the Oxford unabridged dictionary, or Webster's unabridged dictionary. However, we hypothesize that this neologism probably does not refer to the painful consequence of kicking a stationary solid object without protective footgear, but more likely to an unhygienic substance that might collect between unwashed toes. We are in consultation with ethnographic linguists of American culture for further clarification on this fascinating question.

Fundamentally, Dr Gill believes that the fact of Marmite's clear superiority over Vegemite (regardless of the opinion of YYEOs) ultimately rests on the superiority of the source material used in the synthesis of each brand of yeast extract. Marmite ultimately derives from British beers, which are delicious and nutritious, while Vegemite originates in Australian beers, which cause acute intestinal discomfort and painful hangovers when used as directed. Given the evident superiority of British warm bitters over Australian lagers, this conclusion is irrefutable. Moreover, Marmite was the original spread, first marketed by the Marmite Food Company Ltd in 1902[6] – and with experience surely comes perfection. Dr Bolton takes needless exception to this conclusion however. Vegemite, a later imitation YE, was launched in 1922 by a Mr Fred Walker of the Fred Walker Cheese Company in Melbourne, Australia, the same Walker who later became involved in the Kraft-Walker Cheese Company, popularized 'processed cheese' worldwide, and who will thusly go down in history as one of the greatest scoundrels of all time.[7]

Conflict of Interest

Both authors declare that the answer to the questions on your competing-interest form is 'No'. Dr Gill notes however that he has enjoyed the delicious taste of Marmite since a tender age, when his granny used to make him 'Marmite Fingers' for a bedtime snack. Dr Bolton does not own stock in the Marmite company.

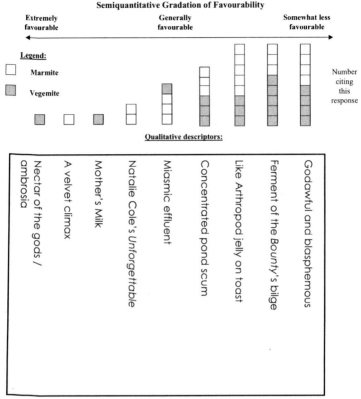

Figure 2

Funding Source
None of the organizations that traditionally support our salaries wish to be mentioned here.

Notes
1. An apparent reference to the aphorisms of one 'T, Mr', an action-film star of some notoriety in the 1980s.
2. That would be Gill, CJ, of 'Gill, CJ et al', occasionally referred to as 'Wise Doctor Gill' as befitting one of his learnedness and insightfulness of opinion.
3. That would be Dr Bolton, a person of Australian origins.
4. See the 2 previous notes.
5. PG Wodehouse's literary character 'Bertie Wooster' offers a convenient guide to its correct pronunciation.
6. www.ilovemarmite.com. A brief history of Marmite.
7. Stradley L, Cook A. *What's Cooking America:* Falcon, 2000.

Yo Mama's *so massive* that ...
Norm Goldblatt, www.normgoldblatt.com

- when she falls down, the Earth speeds up.
- she clogged a black hole.
- her body moves in a straight line at constant speed even when acted on by an external force.
- she can bend light with her bare behind.
- she gave scientists what they've been searching 50 years for — a gravitational wave.
- if she's in the lab, balls roll UP inclined planes.
- when she's in a room, chandeliers don't hang straight.
- she falls at thirty-THREE feet per second per second.
- an 18-wheeler smacked into her, halting the truck and sending her flying 2 miles an hour.
- force equals yo mama's mass times NUTHIN'.
- when she walks down the street, everything in her path gets swept up onto her. That's right. The Astronomical Union recognizes her as a planet.
- she had a fight with a magnet over an iron ingot — and won.
- she INVENTED mass. That's right. She's a Higgs Boson. Shut down the LHC. Not needed any more.
- her head is so big, her center of gravity is directly behind her nose, making for very spectacular breakdancing.
- she can't even moon-walk on the MOON.
- Hooke's Law never applies.
- she broke through the ice — on a rink.
- she sinks in Mercury.
- dermatologists have a hard time removing moles.
- she has to set her Sleep Number mattress to an imaginary number.